Advance Praise for *Reinventing Green Building*

After single-handedly giving us a whole library of excellent books on green building, Jerry Yudelson has delivered his masterwork. If you're going to read one book about what the green building movement must do to address climate change, *Reinventing Green Building* should be your choice.

— Robert Cassidy, Executive Editor, *Building Design & Construction*

Reinventing Green Building offers critical and candid research to illustrate the successes and shortcomings of certification models. He challenges the new professional leaders to re-conceptualize technological solutions to further the revolution that is needed to outpace the direction of the built environment.

— Alison G. Kwok, PhD, Professor of Architecture,
University of Oregon, Eugene, OR

Jerry Yudelson is eminently qualified to write this book. What he has to say is always interesting, often provocative and worth reading.

— Michael Deane, LEED Fellow;
Vice President, Chief Sustainability Officer,
Turner Construction Company, New York, NY

Jerry Yudelson promotes a new approach to sustainability for the 99%. Jerry's strategy is to cut through the bureaucracy and expense of the current rating systems, and to push for simple and profound change.

— Steven A. Straus, PE, CEO, Glumac, Los Angeles, CA

If the 1968 publication of the *The Whole Earth Catalogue* was the birth of the environmental movement, then the challenge posed by Jerry Yudelson in this hard-hitting and clear-minded book is the maturing of global sustainability into adulthood. Bravo Jerry!

— Jonah Cohen, AIA, Principal, Hacker architects, Portland, OR

With wit and courage, Jerry makes a clear-eyed assessment of the flaws of the current sustainable certification systems and lays out pragmatic and inspirational strategies to help us all achieve a better world.

— Bob Shemwell, FAIA, Principal, Overland, San Antonio, TX

In his latest book, Jerry Yudelson lays out a clear case for why a new approach is needed, and presents a practical, well-researched and intelligent path to unlock this potential and generate the results we all strive for.

— Alan Scott, FAIA, LEED Fellow, Director, YR&G, Portland, OR

Sustainable architecture, which today has a high degree of complexity, is not a closed concept but is in a state of continuous development. In his present book, Jerry Yudelson formulates inspiring new ideas on how we have to act to achieve better sustainable results.

— Christoph Ingenhoven, Ingenhoven Architects, Düsseldorf, Germany

Jerry says: "Owners feel that if you're smart enough to be hired to design a building and you're certified as a LEED specialist, you should be smart enough to document their project and give it a final certification. So do I!" I couldn't agree more!

— Richard J. Mitchell, AIA, LEED AP BD+C,
Managing Principal, MACKENZIE, Portland, OR

Either the green building movement will become more successful soon or it will be considered a failure in retrospective. In this context Jerry Yudelson's new book is a great step into this direction; it's a *must read* for everyone who agrees that the built environment must change radically.

— Thomas Auer, Dipl.-Ing., Founding Partner, Transsolar, Stuttgart, Germany

Our fragmented building industry would be well-served by this book which cuts to the chase on challenges confronting the whole business of building delivery and offers an optimistic and pragmatic path.

— Khee Poh Lam, PhD, Professor of Architecture,
Carnegie Mellon University, Pittsburgh, PA

Reinventing Green Building is unique among the hundreds of books about green building. This book asks questions and postulates opportunities for dramatic changes to green building certifications. The timing of this book could not have been better. We've hit the wall.

— Nathan Good, FAIA, Principal, Nathan Good Architects, Salem, OR

Once again Jerry has looked around the corner for us and has the direction we are heading, and it isn't always pretty. A world "scout" in the truest sense of the word, he sees the hazards ahead and creates the necessary markings for a new trail.

— Christian Klehm, LEED Fellow, President,
Energy & Environmental Solutions, Pittsburgh, PA

Jerry knows we must transform the building industry to avoid the worst impacts of climate change. This book makes a strong case for that transition and provides a vision of next-gen green building.

— Mark Wilhelm, Corporate Director, Sustainability and Climate Neutrality
Initiatives, Ameresco, Inc., Phoenix, AZ

Finally, a book that doesn't just tell us how green building ratings are broken, but actually offers real solutions!

— Lawrence (Larry) Clark, GGP, LEED AP, Principal,
Sustainable Performance Solutions, Miami, FL

Jerry Yudelson continues to exhibit a great understanding of the pulse of sustainable design in the built environment. This new contribution will propel the dialog beyond its current entrenchment. *Reinventing Green Buildings* will jump-start the industry to look ahead.

— Ronald van der Veen, AIA, LEED AP, Principal,
NAC Architecture, Seattle, WA

Reinventing Green Building offers critical and straightforward research to demonstrate the accomplishments and deficiencies of certification models. With his analysis, Yudelson challenges the new professional leaders to re-conceptualize technological solutions.

— Thomas Spiegelhalter, Professor of Architecture,
Florida International University, Miami, FL
and Spiegelhalter Architects, Freiburg, Germany

Jerry's ideas in this book are spot-on about disrupting the status quo with a sensible building rating system based on a simple-to-comprehend formula of KPIs driven by smart building technology and data analytics.

— Brad Miller LEED AP BD+C and EBO+M, GGP, GHG-IQ, CDT
President, Environmental Concepts Company, Inc., Irvine, CA

I agree with Jerry's thesis that we need to dramatically improve our utilization of technology and innovative workflows to simplify the deployment of sustainable solutions into the built environment.

— Paul Shahriari, CEO/Founder, ecomedes.com, Atlanta, GA

Through this book Jerry builds a fact-based case on why green building should be the rule rather then the exception, how we can simplify the pathway to get there using new and updated technologies. Simply put, it is a literary masterpiece!

— Beth Holst, former VP Credentialing and Strategic Alliances/
Business Partners for Green Building Certification Institute
and US Green Building Council, Washington, DC.

Jerry Yudelson, a true champion of sustainability, proposes to expand green building certification to 25-percent—far above the one-percent achieved by LEED—using an approach that we should adopt immediately.

— Hernando Miranda, PE, Principal, Soltierra, Inc., Dana Point, CA

Green certifications have arguably fallen short of their promise to transform the real estate industry. Casting a fresh eye on the green building movement, he proposes a smart, simple and sustainable approach to steer us back on course quickly and cost effectively.

— Jiri Skopek, Managing Director, Sustainability,
Jones Lang LaSalle, Chicago, IL

By highlighting issues and identifying solutions, this book will accelerate the conversation to help increase adoption of green building strategies.

— Kelsey Mullen, former director,
LEED for Homes Multifamily Program, Carlsbad, CA

Bringing together the importance of green buildings and the technological challenge with the Internet of Things and big data analytics, Jerry is way ahead of his time. The book is vital to press the reset button and start rethinking of how sustainability can be really integrated into the real estate business.

— Dagmar Hotze, Sustainability Video-journalist, Hamburg, Germany

Jerry Yudelson is the definition of a game changer. In *Reinventing Green Building*, Yudelson focuses his disruptive thinking on a green building market that is thirsty for continued growth.

— Drew Shula, LEED AP BD+C, ID+C, GGP,
Principal, Verdical Group, Pasadena, CA

This book gives an objective, analytic view and constructive critique on how good we were up to now in our market transformation, but at the same time provides a useful orienteering tool for the green community on how to proceed. Jerry has so much to tell. Thanks again for pushing us forward!

— Marija Golubovic, LEED AP, USGBC faculty, Founder and Principal,
Energo Energy Efficiency Engineering, Belgrade, Serbia

Buildings are one of the last big industries that have been essentially unaffected by technology. If our industry is going to really move the needle, then we have to take a dramatically different approach to buildings—all buildings not just our shiny, masthead examples.

— Deborah Noller, CEO and Founder, Switch Automation, Inc.,
San Francisco, CA and Sydney, Australia

Jerry Yudelson is a disruptive innovator in the race to Carbon Neutrality. This call to action is a must-read and a must-implement.

— Sue Sylvester, LEED AP BD+C, Co-Founder,
USGBC Arizona Chapter, Phoenix, AZ

REINVENTING
GREEN BUILDING

WHY CERTIFICATION SYSTEMS
AREN'T WORKING
AND WHAT WE CAN DO ABOUT IT

Jerry Yudelson

new society
PUBLISHERS

Cover design by Diane McIntosh. Cover image © iStock
Printed in Canada. First printing March 2016.

| Funded by the Government of Canada | Financé par le gouvernement du Canada |  |

Paperback ISBN: 978-0-86571-815-9
eISBN: 978-1-55092-611-8

Inquiries regarding requests to reprint all or part of *Reinventing Green Building*
should be addressed to New Society Publishers at the address below.
To order directly from the publishers, please call toll-free (North America)
1-800-567-6772, or order online at www.newsociety.com

Any other inquiries can be directed by mail to:

New Society Publishers
P.O. Box 189, Gabriola Island, BC V0R 1X0, Canada
(250) 247-9737

LIBRARY AND ARCHIVES CANADA CATALOGUING IN PUBLICATION

Yudelson, Jerry, author
Reinventing green building : why certification systems aren't working
and what we can do about it / Jerry Yudelson.

Includes bibliographical references and index.
Issued in print and electronic formats.

ISBN 978-0-86571-815-9 (paperback).--ISBN 978-1-55092-611-8 (ebook)

1. Leadership in Energy and Environmental Design Green Building Rating System.
2. Sustainable buildings--Design and construction.

3. Sustainable construction--Certification. 4. Green technology. I. Title.

| TH880.Y93 2016 | 690.028'6 | C2016-901258-1 |
| | | C2016-901259-X |

New Society Publishers' mission is to publish books that contribute in fundamental ways to building an ecologically sustainable and just society, and to do so with the least possible impact on the environment, in a manner that models this vision. We are committed to doing this not just through education, but through action. The interior pages of our bound books are printed on Forest Stewardship Council®-registered acid-free paper that is 100% post-consumer recycled (100% old growth forest-free), processed chlorine-free, and printed with vegetable-based, low-VOC inks, with covers produced using FSC®-registered stock. New Society also works to reduce its carbon footprint, and purchases carbon offsets based on an annual audit to ensure a carbon neutral footprint. For further information, or to browse our full list of books and purchase securely, visit our website at: www.newsociety.com

Dedicated with great respect and appreciation
to all those men and women, building professionals
of great skill, experience and dedication,
who have given freely of their time and talent
to advance the green building movement
during the past 20 years.

We hope this book leads to a revolution
in our collective thinking and a reinvention
of the art and practice of high-performance building,
based in sustainable cities, for everyone's benefit,
now and in the future.

The future belongs to sustainable cities.

As the Chinese architect Ma Yansong writes
in *Shanshui City* (2015):

If the ancient city concerned religion,
And the modern city concerns capital and power,
Then the city of the future should concern people and nature.

Contents

PART II: GREEN BUILDING HITS THE WALL

Tables

Figures

COLOR IMAGES

Preface

Today's green building rating systems stem directly from pioneering efforts in the UK in the early 1990s when the BREEAM system (Building Research Establishment Environmental Assessment Method) first created what was to become the basic structure for all green building rating systems: broad consideration of key environmental issues, seven to ten broad categories weighted by relative importance, and a point scale for tallying up achievements and rating levels based on overall point totals. The intended result? Green buildings and a better environment! The result in practice? The subject of this book.

The core argument for green building remains: as a society we urgently need to address climate change issues, and buildings are responsible directly (via energy used in operations) and indirectly (via product manufacture, employee commuting, etc.) for more than 30 to 40 percent of global carbon emissions. This alone argues for green building systems that are tightly focused on reducing carbon emissions through a full life-cycle analysis of building construction and operations.

In 1997, I first heard the term "green building" from two people: my boss, Steven Straus, CEO at Glumac engineers in Portland, Oregon, and my friend Nathan Good, FAIA, an award-winning green architect (Nathan and I co-founded Green Building Services in 2000, an early consultancy in the field). At the time I had a two-decade background in energy and environmental affairs, so to me the term made perfect sense. Why shouldn't buildings be designed and operated to have fewer environmental impacts? I went to my first US Green Building Council (USGBC) national meeting in 1999 and became a USGBC board member in 2000. In 2001, I became one of LEED's first ten trainers in the United States and one of the first hundred LEED Accredited Professionals. Also in 2001, I helped certify the second-ever LEED Gold (version 2.0) project, the Jean Vollum Natural Capital Center in Portland, Oregon. In 2011, the USGBC and the Green Building Certification Institute elected me to the first class of LEED Fellows.

To help accelerate the market transformation toward green building, I began writing books for professionals in the field. In 2007, I wrote *The Green Building Revolution*, outlining why I thought the time had come for rapid growth in green building. In the same year, I wrote *Green Building A to Z*, which defined the key terminology for accomplishing this growth, and also wrote *Marketing Green Building Services*, to show professional service firms and contractors the advantages in addressing this emerging market.

Beginning in 2000, I directed a number of LEED certification projects. The most recent (in 2012) took more than a year to go through the certification process. Even as a long-time market participant and advocate for LEED, it became obvious more than five years ago that the system did not meet many marketplace needs.

In 2014 and into 2015, I headed the Green Building Initiative (GBI), USGBC's only current competitor in the US market. At GBI, I learned another approach to green building: one that relied on a national consensus standard (which LEED is not) arrived at through a well-recognized process—a rating system called Green Globes that utilizes trained and certified assessors to deliver the judgment about a project onsite to the building team; one that is often much cheaper and faster than LEED in delivering results.

I'm a professional engineer by training and a marketer by inclination, with many years of business experience. I respect both technical concerns and marketplace judgments. As I studied alternatives, I thought: *Why couldn't we have both: a technically valid approach to green building, but cheaper by a factor of 10 (or even 100)?*

Then I thought: if we could create a very inexpensive green building rating system, yielding real benefits quickly, people would fall all over themselves to use it! That thinking process led to writing this book.

I hope that a new generation of tech-savvy green building advocates, not wedded to outmoded concepts and certification models, will pick up the innovation baton (which has been dropped by USGBC, in my opinion) and get the green building revolution moving forward again, at the rapid pace that the planet's health requires.

— Jerry Yudelson

Foreword

As a lifelong environmental advocate, and a green building consultant for the past 20 years, I was thrilled to be involved in the green building movement's meteoric growth. But more recently I have been concerned about where we find ourselves today.

With the impacts of climate change, both subtle and profound, becoming clearer every year, we need an approach for radical climate action that will be rapidly and broadly adopted in buildings. We need a way to bring us back from the brink of climate chaos, quickly and cost-effectively.

The US Green Building Council (USGBC) has responded to the threat we face by modifying the Leadership in Energy & Environmental Design (LEED) standard to place more emphasis on reducing energy and climate emissions, most recently with LEEDv4. At the same time, LEED is losing its momentum and becoming a tool embraced by a relatively small and diminishing segment of the industry.

While many have argued that the USGBC should simplify its approach and engage a broader segment of the market, LEED's strategy has instead subtly shifted from market transformation to leadership recognition, in a way that is leaving the industry mostly "LEEDerless."

Getting a smaller and smaller segment of the population to jump higher and higher is not going to get us where we need to go.

While early adopters, market leaders and the truly committed are embracing the higher thresholds of LEEDv4 and exploring even more demanding standards such as the Living Building Challenge, Passive House and zero net buildings, the great majority of real estate owners and practitioners are keeping their heads firmly planted in the sand. This is simply unacceptable, since buildings are responsible directly and indirectly for almost 50 percent of global climate emissions.

It is time to take a hard look at what is working and what is not and take a zero-based approach to what we must do next to engage the broader real estate community in the fight against climate change.

This book and its author, Jerry Yudelson, do just that, clearly and succinctly describing not only what is wrong with what we have today but also clarifying why we should feel hopeful. There is a path to a post-carbon world where building energy use has been slashed and renewable energy sources provide the majority of our heating, cooling and electricity, but it's probably not the path we're on. The book suggests that we may need to start again or seriously rethink our strategy with the end clearly in mind.

For Yudelson, "begin with the end in mind" dictates a laser-like focus first on carbon reductions and then on water use. If we must de-emphasize other sustainability goals in the short term to create an easier, less daunting path for real estate to reduce carbon emissions, it is a price we must pay.

He also points the way to how we can simplify and automate the process, thereby reducing costs dramatically. Some organizations give lip service to algorithms, "Big Data," and the "Internet of Things" but no one has really tapped into the power of modern technology and computing to deliver, and more importantly to document, green buildings.

The mind-numbing detail and pencil pushing that is driving practitioners crazy and costs through the roof would be more appropriately addressed by computers with advanced algorithms, using data that is already contained in computer-based design documents or building management tools. The author makes the case that green building certification can be easy, inexpensive and automatic, if it is integrated into the tools we already use for design and operations management.

"Smart, simple and sustainable" sounds like a great design brief to me. There is enormous opportunity for innovation and entrepreneurial efforts to connect existing data to a largely automated system for green building guidance and certification. This book is a must read for anyone who is interested in delivering on that promise.

Who else should read this book?

- Everyone engaged with existing green building certification systems, especially the leadership of the USGBC and Green Business Certification Inc. (GBCI). As the author details, LEED has achieved many accomplishments and established itself strongly in the field. It would be a tragedy if they were unable to recognize the need to change and to nimbly shift direction.
- People worrying about the state of green building who can't yet see their way to a different approach to transforming the marketplace in the face of climate change.
- Anyone who has been curious about the increasing number of green building tools that are currently available and their differences, strengths and weaknesses.
- Decision-makers at all levels of government who are tasked with achieving significant reductions in greenhouse gas emissions or water use.
- Anyone who is interested in current trends in cloud-based technology and data platforms and how they can simplify green building certification.
- Everyone in real estate who is concerned about climate change and wants to understand their important role as leaders and consumers of transformational technology and services. If we are to ultimately win this battle, real estate must transform itself from a massive contributor to global environmental problems to an important part of the solution.

Jerry Yudelson has the experience and the vision to point the way, but success will be defined by how many people read this book and take its lessons to the marketplace. It is insightful, timely, strategically important and a fascinating topic. I encourage everyone to read this book and look forward to continuing the conversation.

— Pamela Lippe, LEED Fellow
Principal, e4, Inc.
New York, NY

Reinventing Green Building: A Call to Action

The major US green building rating system, LEED, isn't growing; the green building revolution has stalled; no easy solutions are in sight. By 2015, LEED had certified *less than one percent* of commercial buildings and homes in the United States during its first 15 years. Annual project registrations and certifications for LEED in the United States are now fewer in number in 2015 than in 2010. It's time for a new green building program that works for "the other 99 percent" and that has significant annual growth.

Figure I.1 compares LEED certifications with the total number of US commercial buildings; it shows that total project certifications at year-end 2015 amounted to *less than one percent of the US nonresidential building stock*. (In the residential sphere, the fraction was considerably less.)[1]

We need a new way to rate buildings for their climate and environmental impacts. As the leading green building organization and largest rating system in the United States as well as the largest in the world, the US Green Building Council and LEED have a special responsibility to engage in self-criticism and continuous improvement.

These concerns are not new, but they have taken on more urgency with the upcoming mandatory switch to a new version of LEED (LEEDv4) in October 2016. With most project teams content in knowing how to navigate through LEED 2009, despite its costs and complexities, LEEDv4 appears to be "a bug looking for a windshield."

That LEED is broken is not news; Randy Udall and Auden Schendler first raised the issue in 2005 with a provocative article,

Total number=
5,500,000

Green Building certified =
34,000 (end of 2015)

0.7%

FIGURE I.1. Total US Building Stock vs. LEED Certifications, End of 2015.

"LEED is Broken—Let's Fix It."[2] At the time, many LEED advocates, including me, dismissed issues raised by this article as simply a reflection of growing pains for the LEED system. At the time, LEED was barely five years old and just getting started on the road to dominating the US market for commercial green buildings.

But their five main objections—LEED is too costly, project teams are too focused on gaining points and not on results that matter, LEED's energy modeling is fiendishly difficult, LEED's bureaucracy is crippling and LEED's advocates continually produce overblown benefit claims—remain drawbacks today.

Most experienced green building professionals would also agree that these same issues remain relevant in 2015. But there is a larger problem: Green building rating systems have diverged greatly from building owners' and operators' core concerns, as these systems are designed to meet the needs of green idealists more than those of most market participants.

Green building advocates must abandon the approach they have taken for the past 25 years: comprehensive and overly technical criteria, multiple elaborate rating systems, large and cumbersome bureaucracies, high costs and inadequate focus on real long-term building performance. Instead, they need to embrace the technological revolution that has cut costs for communications by factors of not ten, not one hundred, but a thousand or more in the past 15 years.

Moore's Law, first enunciated in 1965, says that computing power doubles every 18 months; over time, unit costs for computing have fallen in a similar fashion.[3] Consider this: Every six years, it's 16 times cheaper (and faster) to do the same task, every nine years 64 times cheaper! (Every 15 years, it's 16 × 64, or 1,024 times cheaper!) With the advent of mobile communications, social networks, the Internet of Things, Big Data analytics, cloud computing and global information systems, why should green building still be governed by concepts, systems and procedures developed in the 1990s "Dark Ages" of Internet 1.0?

This book's central thesis is that it's time for a serious debate about LEED's (and other systems') inadequacies in addressing a few key issues: combatting global climate change, addressing looming water scarcities and reducing resource waste.

The corollary is that it's time for green building leaders to develop a new model for certifying project design, construction and operations, one that is:

- **Smart**: technology-savvy and mobile-accessible
- **Simple**: so anyone can understand green building standards without specialized training and certification
- **Sustainable**: both in focusing on absolute performance as the best means for addressing climate change, and in accelerating building design and management's movement onto cloud-based platforms.

We don't need to abandon concerns about urban design, healthy buildings, or healthy building materials—but they belong in a separate system or systems. Future green building rating systems should focus ONLY on five Key Performance Indicators:

- Energy use
- Total carbon emissions
- Water use
- Waste minimization
- Ecological purchasing

Until we build most new buildings and retrofit most existing buildings according to dramatically higher standards for energy, carbon, water, waste generation and recycling, then all other considerations are window dressing.

After all, Nature doesn't care how much we *reduce* annual carbon emissions from unsustainably high levels. Nature only cares about *absolute* levels of carbon dioxide (and other greenhouse gases) in the atmosphere, about excessive water use that damages natural ecosystems and about waste that doesn't get recycled into something else.

It turns out that the solution is already staring us in the face: the technological revolution that has given us the mobile Internet, social media and Big Data analytics. With this revolution, we can start with the user's concerns and work toward creating a rating system (or systems) that enhances the user's experience.

How to proceed? Here's an example in one word: Uber.

In 2015, just five years after it started, Uber's latest financing round valued it at $50 billion. What did Uber do? It took on a hundred-year-old urban transportation system—taxicabs—and created an easy-to-use smartphone app that revolutionized it, in the process challenging and upending a highly regulated, low-user-satisfaction industry.[4] No one likes taxis, but if you land at any airport or stand on any street corner in any large city, they're usually the only curb-to-door service available.

What don't we like about taxis? They're not always available when and where you want them; they're hard to get during rush hour, rainstorms and at dinnertime; they are often dirty and uncomfortable; they are prone to occasional customer rip-offs; and they may not accept credit cards for payment. The taxi business's main beneficiaries are taxicab owners, not customers or even drivers.

Uber started with the idea that a ride-for-hire service could address these issues, utilize surplus labor and vehicles, enhance customer experiences and be profitable for all concerned—by using the phone we already carry in our pockets. Brilliant! I've used Uber's smartphone app many times: I can track where the driver is at all times; I know I'm going to get a clean and comfortable car with a driver who knows the town; and I've already paid the fare and tip when I step into the vehicle.

Uber is so disruptive that it has encountered stiff opposition from everyone profiting from the current system, including "progressive" politicians who are in hock to taxicab owners for campaign contributions, but it will succeed because it's focused on creating a superb user experience. By one account, nearly two million New York City residents have already downloaded the Uber app![5]

Green building certification is ripe for the same disruptive treatment, but it's supremely unlikely that established organizations can or will upend their current revenue models to provide a far more user-friendly approach. It's time for new organizations and fresh thinking in green building. It's time to leave behind the current monastic, hair-shirt experience of LEED certification and create a fabulous user experience. In short, it's (past) time for *Reinventing Green Building*!

THE GREEN BUILDING MOVEMENT

CHAPTER 1

The Technological Challenge:
The Age of Algorithms and Big Data

Once a new technology rolls over you,
if you're not part of the steamroller,
you're part of the road.

Stewart Brand[1]

Prepare yourself for the coming era of ubiquitous and never-ending connectivity. Since the 2001 introduction of the iPod, digital innovations are numerous: the iPhone in 2006, the iPad in 2010, the Apple Watch in 2015, mobile computing's global growth, billions of people on social networks and social media, the advent of the sharing economy, the rise of cloud computing (e.g., via Amazon Web Services, Microsoft Azure and Apple iCloud) and the growing "Internet of Things" (IoT) connecting every imaginable digital device, *all in real time.*

According to Dr. Osman Ahmed at Siemens, four major technologies are "game changers" for anyone in the business of owning or managing buildings:

1. The Internet of Things (IoT), comprising billions of now ubiquitous connected data-gathering devices (an estimated 50 billion interconnected devices by 2020). The IoT is a new market potentially worth $70–$150 billion by 2025.[2]

2. The cloud, which makes computing power available to anyone, anywhere, from any connected device.

3. Mobile devices and open systems, which allow everyone with a smartphone to have access to thousands of apps that help them to manage their life.

4. Big Data analytics, which has the computing power to handle millions of data points from diverse and disparate sources and provide prediction and diagnostics for better managing building operations and energy use, coupled with algorithms guiding use of Big Data.[3]

To this list, we can add widely available cheap sensors that flow building operating data up through the IoT, the cloud and Big Data onto mobile devices. Taken together, these five technologies represent the key trend driving the market for automating building operations (Figure 1.1).

Industry expert Realcomm says that standardized, flexible, secure and state-of-the-art IP networks (both fixed-cable and wireless) make tremendous sense to put in each and every green building, old and new, *right now*, for multiple reasons:[4]

- **Cost**: running many different wiring systems up a building's spine is costly.
- **Adding New Systems**: it's easier and less costly to use a standardized approach.
- **Integration**: without a standard network, getting many different systems to talk with each other is both time-consuming and costly.
- **Support**: supporting the networking/communication needs for many different systems can be difficult and as well as costly.
- **Management Ease**: multiple systems are harder to manage.
- **Security**: securing and managing a single IP network architecture is easier.

In a few years, fully capable building IP networks will become as common as mechanical and electrical systems are today. Quite soon, all equipment that goes into buildings will be IP-ready and based on open systems, making this transition even easier. According to one expert, the IoT model for buildings breaks down into seven functional levels:

Devices are connected and send and receive data interacting with the *Network* where the data is transmitted, normalized

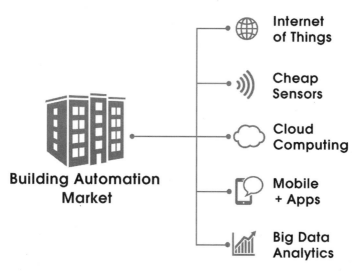

FIGURE 1.1. Five Technologies Driving the Market for Building Automation

and filtered using *Edge Computing* before landing in *Data storage* and *Databases* accessible by *Applications* which process it and provide it to *People* who will act and collaborate using the data.[5]

Building design, construction and operations are about to undergo seismic shifts, with advanced design software and cloud-based building management systems. Yet during the past 15 years green building certification has hardly changed, except that (in some cases) data for certification can now be extracted from design documents and operational data, and then uploaded to web-based platforms. It is, however, still evaluated item by item by review teams of professional consultants and building assessors.

There are a few exceptions. Paul Shahriari, an experienced software developer for green products, created a decision-making app, *ecomedes*, that allows one to easily calculate the payback from green investments such as water-conserving toilets and fixtures (Figure 1.2).[6]

In 2012, USGBC introduced an app, the Green Building Information Gateway (GBIG) that allows you to find on your smartphone LEED-certified green buildings in any city, using USGBC's project

FIGURE 1.2. The *ecomedes* App. Credit: ecomedes, LLC, Paul Shahriari

database.[7] That's just about the extent of the mobile revolution in green building.

Just as mobile and cloud-based technologies have totally disrupted revenues and business models for such technology leaders of the 1990s and 2000s as Intel, Dell, HP, Nokia, Microsoft, Research in Motion, etc., and made Apple the most valuable stock of any company in the world in 2015, isn't it logical to expect that these disruptive technologies will soon challenge and disrupt green building design, construction and operations as well as certification's current analog-based model?

In *The Attacker's Advantage*, management guru Ram Charan writes about the "structural uncertainty" of our times:

Every day more [people] have instant access to any and all knowledge and insights that exist, as well as the ability to collaborate with others as never before. Their ideas can be scaled up swiftly, because capital is readily available to fund promising ideas. For digital companies, the scaling can be accomplished extremely fast and at low incremental cost. On the other side of the coin, consumers have acquired great new powers because of digitization and online connectivity...that give them information and options they never had before.… Every uncertainty is magnified by the quantum increases in the speed of change.[8]

This book's central thesis is that these rapid technological changes present a major challenge to our current approach to green building certification, but also represent significant opportunities for making changes that could greatly expand the market for green building by lowering costs dramatically and engaging both building owners and occupants in reducing energy use. We'll return to this theme often in this book.

The Great Convergence:
Real Estate, IT, Energy and Sustainability

Given technology's constant and rapid change, what's happening with the built environment and building operations? How has it responded to digitization and online connectivity? Neglecting for a moment the revolutions in architectural design and building construction, it's easy to see how this revolution is reordering building operations. Figure 1.3 shows this convergence.

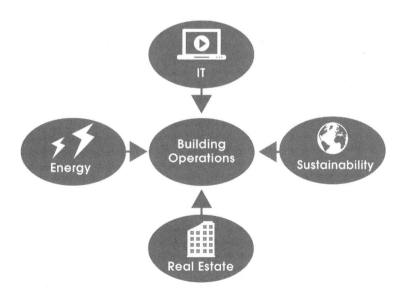

FIGURE 1.3. The Convergence of Real Estate, IT, Energy and Sustainability

Essentially, information technology and real estate have been converging for many years. For example, in large swaths of commercial real estate, most rents are billed and collected online. What's new since about 2011 is that the convergence has added energy and sustainability as key components.

New digital "dashboards" allow near-real-time assessment of a building's electricity use, typically in 15-minute intervals, compared against (an engineer would say "normalized to") any parameters you want: climate zone, local weather, occupancy, building type/use and geographic location; and display that information in many visual formats, readily available on any Internet-connected device to anyone who has access rights. In a building, as shown in Figure C.12 (color plates), one could display for example, daily, weekly and monthly building energy use to tenants or occupants as a way to influence behavior or reward good efforts.

Dashboards and similar cloud-based software can also handle hundreds of buildings simultaneously, allowing portfolio managers and third parties to readily observe and act on information in real time, since electricity use is time-stamped daily and even hourly. Because electricity use (along with monthly gas, diesel, water, etc.) can be easily analyzed, visualized, communicated to anyone who needs to know via SMS (text) message and reported in various formats, it should be easy to assess a building's key sustainability performance variables and to improve them using these new management tools.

Of course, that's good for assessing operations for existing buildings, but what about new construction? Here again, the digital revolution is gaining speed, with Revit and similar BIM (Building Information Modeling) software used for design, from which it should be possible to perform a complete green building assessment with very little user or assessor interaction, knowing building location, design features and product selection.

How hard would it be to write a piece of software that looked at the floor-by-floor layout of a building and determined whether 90 percent of the building occupants had a view of the outdoors or had adequate lighting levels, and then to connect that information to a green building rating system? Why should anyone be tasked with documenting or assessing such a feature when an algorithm could do it so much easier and faster? Why are the leading green building rating systems not moving to create APIs (Application Program Interfaces) that third parties could use to create these features?

Paul Shahriari started *ecomedes* to address a critical problem, one that many market leaders are ignoring:

> Somebody has to create linkages between component product manufacturers' data and the characteristics of those products as a whole, and then somebody at the automated design platform needs to specify—how many square feet do we need? [But the problem is] with sustainability, since to get more detail out of our projects we are dramatically increasing the amount of work that has to be done by a project team. That is not a good recipe for growing adoption and a market. To me, solving that is one of the first steps, and then all the other stuff can also be done digitally.[9]

The Age of Algorithms

The entire green building assessment process could easily be turned over to algorithms, eliminating a need for third-party auditors or assessors, as well as expensive consultants, essentially allowing projects to "self-certify" and cutting costs by 10 or even 100 times.[10] Every project that met certain criteria could be labeled a "green" building, without engaging consultants, requiring uploads to a website by dozens of hands, review by "experts," and finally, many months (even a year or more) later, receiving a rating at a certain level, as is now the case with LEED certification.

In later chapters, we will explore what this opportunity will require, e.g., eliminating "analog" criteria currently embodied in most green building rating systems and substituting strictly "digital" criteria. But for now, consider why a focus on algorithms and digital assessment is so important.

Ram Charan comments:

> The single greatest instrument of change…is the advancement of mathematical tools called algorithms and their related sophisticated software. Never before has so much mental power been computerized and made available to so many.… In combination with other technological factors, algorithms are dramatically changing both the structure of the global economy and the lifestyles of individual people.[11]

We already know that "flash trading" on Wall Street, done strictly with algorithms, has made it virtually impossible for smaller and short-term investors to gain any advantage.[12] Algorithms are far smarter than any individual, far quicker to apply and far more able to incorporate new information as fast as it becomes available.

Bringing Green Building Certification into the Age of Big Data

With such dramatic changes underway and more on the horizon, why is green building, including its certification protocols, still stuck in the "horse and buggy" stage? As we will demonstrate later, one consequence is that, more and more, building owners are turning their backs on certification and leaving certification's "benefits" on the table, as a reaction to high costs, long certification times and the system's complexity.

The next big problem is trying to assess product "ingredients" on the numerous products specified for a typical building. Try sorting through Environmental Product Declarations (EPDs) and Health Product Declarations (HPDs) by the hundreds (soon to be thousands), without using advanced digital tools, and you will see the colossal waste of time and money embodied in our current approach to assessing building product "greenness," one spec sheet at a time.

Fortunately, new tools keep evolving to meet this challenge. Only cloud-based software can help you make sense of thousands of products and building systems' EPDs using "data sheets" from each individual product, which manufacturers constantly change. Without using cutting-edge software to automate creating and delivering environmental performance assessments and material health evaluations, in a way that provides understandable and meaningful information, says Sustainable Minds' CEO Terry Swack, there is no real benefit to having the information:

> Today's EPDs can't help make greener purchase decisions— other than enabling a box to be checked [on a LEED form]. There's little to no information in an EPD that an architect, builder, contractor can really use to say, "Hey, this actually is a greener product, I'm going to buy it." Other than industry average EPDs, there are no LCA (Life-Cycle Assessment) product benchmarks, and EPDs for similar products from different manufacturers cannot be compared for a variety of reasons.[13] The only requirement [currently] is to deliver the technical data. There is no requirement to help specifiers figure out how to use it.

According to Swack, manufacturers should build their brands by providing credible and *understandable* environmental information about their products. In this way, they can over time build a greener brand by *demonstrating they know what they're doing.* Users who want to make greener purchase decisions require consistent, understandable and meaningful information.[14]

A Smorgasbord from the Internet of Things

Many technologies on the market today collect information from building operations; analyze it; visualize it; and tell someone what to do about it, how to control it better and communicate it to building occupants. The following examples showcase a few representative technologies that green building advocates should know about and use to change how they approach greening the built environment. These technologies exemplify a revolution that is making everything *Smaller, Faster, Lighter, Denser, Cheaper*[15]—helping us to realize how to create sustainable changes in buildings by taking a ride on the information revolution's bullet train.

Wireless Sensors. In 2011, I gave a talk in Toronto to the EnOcean Alliance, a trade group that develops standards for wireless sensors. In essence, since every building has wireless connectivity (as do most homes) why not have wireless sensors everywhere? They're easy and cheap to install and work via electricity stored in capacitors that capture several forms of ambient energy: photoelectric (from ambient lighting), a phenomenon first explained in 1905 by Albert Einstein; thermoelectric (working on only 2°F temperature differences), a technique that has powered devices for decades; and the piezoelectric effect, where pushing an on-off switch can transfer that energy into electricity via an organic "crystal."

What's the implication? In a hotel for example, every room could be equipped with a key card that would turn off all the HVAC equipment and lights whenever you left the room, saving considerable energy. It could also be controlled from the front desk, so the air-conditioning or heating could be turned on to your desired temperature before you check in. This opportunity is technologically available right now; according to experts, in 30 minutes one can retrofit a hotel room with a key card, using "peel-and-stick" sensors.

Wireless sensors can connect back directly to certification systems, as can sensors in intelligent lighting systems. It's exciting to think about all the data that will soon be at the fingertips of building owners (and cheaply too)!

Cheap Electrical Submeters. Coming to the market soon will be a stick-on electrical submeter, a wireless sensor that attaches directly to an electrical line coming into an electric breaker at a panel, reads electricity use, sends data out wirelessly and powers itself using just line current.[16] You can submeter everything quite cheaply, collect the data wirelessly and analyze it on a Big Data platform, really getting a handle on electricity consumption by end-use. Pretty cool stuff!

Remote Building Audits. Consider again what software and algorithms can do to make data analysis for green building operations readily accessible and affordable. A company like FirstFuel Software can provide building audits and actionable recommendations for energy efficiency upgrades by combining a building's electricity meter and weather data with third-party data, building information and their own building engineering expertise. Loaded into the company's platform, algorithms can deliver actionable insights for every building. The company then applies its own deep-data frameworks, along with advice from experienced building scientists and engineers, to provide energy conservation recommendations, performance monitoring and alerts, forecasts and energy savings estimates—all tailored to each specific building.[17]

Using algorithms to analyze large amounts of data in the cloud in real-time can help companies and organizations tackle annual energy waste amounting to tens of billions! As an example, working with the US General Services Administration and "analyzing the 4.1 million-sq. ft. Ronald Reagan Building in Washington, DC, FirstFuel found that two large exhaust fans were unnecessarily operating at full speed. Adjusting the fans' set points to their original design levels contributed to the Reagan building saving $800,000 in one year."[18] Overall, GSA audited 180 buildings remotely using building data and the FirstFuel algorithm, and found $13 million in annual savings. This shows software's power to guide users toward energy-saving measures quickly and accurately, without expenses for onsite audits, engineering studies and consultants' reports.

Comfort via Mobile App. Another app called *Comfy* uses a smartphone to allow a building occupant to "dial up" more heating or cooling in whatever space they happen to be in, using either a mobile device or a desktop computer (Figure 1.4).[19] More importantly, the app learns over time what temperature is best at any location in the building. Empty spaces would be conditioned less, saving energy. Key to its value is that user engagement helps to improve productivity and accommodate a workforce that has diverse comfort requirements.[20] And, as a bonus, there may be no more intra-office fighting over the thermostat setting!

FIGURE 1.4. Comfy App Allows Users to Request Exact Temperature for Their Comfort. Credit: Courtesy of Building Robotics, Inc.

Automated Energy Star Reporting. Several building dashboards currently can automatically report energy use data to Energy Star and qualify a building for an Energy Star score and, if it scores in the top quartile of similar buildings, secure recognition as an "Energy Star" building from the US Environmental Protection Agency. (See Chapter 12 for further discussion about Energy Star.) Lucid's BuildingOS platform, shown in Figure 1.5, provides this service for building portfolios.[21] This process also allows easy real-time analysis of a portfolio's performance and lets managers immediately identify poorer performers and make plans to upgrade them.

The future will get even more engaging, as green building pioneer John Picard predicts,

> Imagine that the skin, security, energy, water, controls, mechanical, lighting and occupants of a building are all talking to each other in real time. The buildings are sensate, adaptive, regenerative, cost effective and healthy. It's a reality on the very near horizon…the availability of data will guarantee more green buildings.[22]

FIGURE 1.5. Automated Energy Star Reporting With Lucid's BuildingOS Platform. Credit: © Lucid

Summary

These are just a few examples of how technological changes, including mobile, wireless, Big Data and cloud-based technologies, along with the age of algorithms, have arrived at the doorstep of building owning and managing. With this much technology already available to manage sustainability in buildings, it's surprising that green building certification does not just piggyback on such platforms to cut costs and dramatically improve user satisfaction.

In the next chapter, we'll look at "Megatrends" in the built environment that are changing how buildings are built, operated and managed to produce green outcomes. These megatrends can become drivers for changes in our current green building certification systems and they help us understand how to "reinvent green building" by using vastly increased capabilities provided by new technology to make everything smaller, faster, lighter, denser and cheaper.

CHAPTER 2

The Future of Green Building:
Top Ten Megatrends

*I became interested in long-term trends because an
invention has to make sense in the world in which it is finished,
not the world in which it is started.*

Ray Kurzweil[1]

In 1982, futurist John Naisbitt introduced "Megatrends," i.e., large-scale trends building on such immutable facts as the unprecedented number of Baby Boomers born in the two decades after World War II.[2] In 2007, Patricia Aburdene updated the concept in her book *Megatrends 2010: The Rise of Conscious Capitalism,*[3] with its focus on the growing interest in corporate social responsibility, a trend clearly driving sustainability thinking and green building forward.

What megatrends currently affect green buildings? (See Figure C.1.) In this chapter, you'll learn about ten megatrends that will shape and drive green building technologies, markets, government rules and certification systems through 2020 and beyond.

A megatrend's strength is that we can't wish it away; it's here to stay and the key issue is how to build on it. In technology, the turn to mobile computing after the introduction of the iPhone in 2007 and the iPad in 2010[4] is clearly such a trend, one that building owners and green building proponents increasingly must deal with.

Globally, green building will likely continue its growth, especially considering green building's rapid uptake in countries in the Asia-Pacific region, South America and the Middle East. Each year, more government agencies, universities, property developers and corporate real estate managers incorporate green design ideas and measures into their buildings and facilities, and there is nothing on the horizon that will stop this megatrend.

The top ten megatrends affecting green buildings:

1. Green building certification's growth rate is flat in the United States.
2. Energy efficiency leads the way.
3. Zero net energy buildings (ZNEBs) are on the rise.
4. Competition among rating systems will increase.
5. A sharper focus on existing buildings will emerge.
6. Cloud computing and Big Data analytics will provide much needed direction.
7. Cities and states will demand building performance disclosure.
8. Debate over healthy building materials will become even more vexatious.
9. Solar power will finally break through.
10. Expect a heightened emphasis on water conservation.

Note that five of these ten megatrends revolve around energy: energy efficiency, zero net energy, cloud-based (and data-driven) energy management, energy performance disclosure and solar power. This focus is largely driven by two practical considerations: first, for most buildings, energy is the largest uncontrollable operating cost; and second, growing understanding about the connection between building energy use and global climate change means that corporate social responsibility and government action are also key drivers for improving building energy efficiency.

Although green building originally focused on the "triple bottom line" (energy; economy, environment and equity; social well-being), concern for energy issues increasingly is driving corporate and governmental interest in green building.

Megatrend #1: Green Building Certification's Growth Rate Is Flat in the United States

Green building in North America, Europe, the Middle East and Asia-Pacific will continue to grow, but at a slower rate, as more building owners come to accept the business case, especially in larger office buildings, corporate real estate, high-end universities and state and federal government buildings. The discussion following Table 4.1 estimates that green building project registrations accounted for about 15 percent of total new construction project area in the United States during 2014.

In contrast to the new construction market, the 2014 tally of 185 million sq. ft. in existing building certifications (Table 4.1) represents only 0.2 percent of existing US commercial building area. At the end of 2015, the US total for LEED certifications is still quite low—about 3.2 billion sq. ft., or about 3.8 percent of commercial buildings' total 85 billion sq. ft. (Figure 4.2).

Slower growth in green building certification in new construction doesn't mean that designers are ignoring important elements of sustainable design, but this book's thesis is that without major changes certification in the United States will show no growth, with important implications both for certification organizations and for sustainability in the built environment.

Megatrend #2: Energy Efficiency Leads the Way

Beginning in 2012, energy-efficiency green building retrofits have shown stronger growth than energy-efficient new construction. This trend is strongest in corporate and commercial real estate, along with "MUSH" market (Municipal, University, School and Hospital) projects. In the MUSH market, cheap capital and financing from ESCOs—Energy Service Companies—drive building owners to "sell" most future energy savings in exchange for investors providing an upgraded or modernized physical plant. (The federal government equivalent, Energy Service Performance Contracts, chose 16 national vendors in 2012 to provide projects worth $85 billion to government agencies.)[5] In *The World's Greenest Buildings: Promise vs. Performance in Sustainable Design*, I make a case that *absolute* building performance, with resultant lower operating costs (vs. the currently more common "relative improvement" approach), should be the primary focus for green building if we want to rapidly cut carbon emissions.

There are huge opportunities in energy efficiency and most are concentrated in 25 percent of the building stock, as shown in Figure C.15. A more cost-effective approach to certifying existing buildings should first make use of the concentrated nature of these efficiency opportunities by launching a rating system tailored for such buildings and their key performance indicators (Chapters 14 and 15 provide a more extended discussion of this opportunity).

Megatrend #3: Zero Net Energy Buildings Are on the Rise

Zero net energy buildings are increasingly commonplace (Figure 2.1). A 2014 survey by the New Buildings Institute (NBI) identified more than 160 ZNEBs in the United States, with an additional 53 low-energy buildings that were "zero net energy capable."[6] Commercial real estate developers (and in some places, new home developers) have also begun to showcase zero net energy designs to differentiate their projects, as have some developers wanting to upgrade older

FIGURE 2.1. Zero Net Energy at the Olver Transit Center, Greenfield, MA. Powered by a 98 kW PV array, geothermal heat pumps and a wood pellet boiler, this 24,000 sq. ft. transit center in western Massachusetts operates on only 32,000 Btu/sq. ft./year.[7] Serving 70 occupants and 100 visitors per day and completed in 2012, it is an early example of zero net energy architecture in a difficult climate. Credit: John Linden

office buildings (see Chapter 14). This trend developed slowly after starting in about 2011 and now seems ready for takeoff.

Megatrend #4: Competition Among Rating Systems Will Increase

In US new construction ratings, LEED may see enhanced competition from Green Globes and possibly from new entrants in specialized niches, e.g., retail or office interiors. (See discussion in Chapter 12.) In 2013 and 2014, the federal government put both LEED and Green Globes on an equal footing for government projects, lending further legitimacy to Green Globes. Other North American systems (Figure 2.2) include LEED Canada; BOMA BEST (from BOMA Canada), which is focused on the existing building market; the Living Building Challenge; and BREEAM.

In Europe, the BREEAM rating system is aggressively marketing itself, particularly in Western Europe, where it competes with country-specific systems such as HQE in France and DGNB in

FIGURE 2.2. North American Green Building Certification Systems

Germany and Austria. In all, BREEAM International is marketing the system in nearly 60 countries and may even enter the US market should the USGBC's new LEEDv4 falter; BREEAM International has already entered the market in Mexico.[8]

In Asia-Pacific, a more likely scenario is for country-specific rating systems to be dominant, especially in more established markets such as Australia, Singapore, Japan, India and China. In Canada, LEED Canada competes in the existing buildings market with BOMA Canada's BOMA BEST rating system.[9]

Megatrend #5: A Sharper Focus on Existing Buildings Will Emerge

Beginning with the Global Financial Crisis (or Great Recession in the United States) in 2008–2010, the green building industry began to switch from evaluating new building projects to assessing existing buildings and tenant spaces. This trend crested in 2009 (Figure 8.8), but it could re-emerge for two reasons. First, third-party green building project registrations for new construction peaked during 2012–2014 and is now a steady 2000–2500 projects per year in the United States, representing annually about 275 million sq. ft. of new building construction (Table 4.1 and Figure 4.4).

With annual new construction registrations for LEED essentially flat, the existing building market may get greater attention, particularly with energy efficiency retrofits and a renewed focus on using the Energy Star system. But LEED may not benefit from this trend. LEED certification is not as newsworthy as it once was and LEED existing building certifications accounted for fewer than 545 buildings in 2014 and 584 in 2015, in each year representing about 0.01 percent (that's not a typo!) of the total US nonresidential building stock of 5.5 million buildings.

Megatrend #6: Cloud Computing and Big Data Analytics Will Provide Much Needed Direction

As we discussed in Chapter 1, building owners and third-party service companies increasingly manage larger buildings remotely, using software platforms that provide performance monitoring, data analytics, visualization, fault detection and diagnostics, portfolio energy management and text messaging, all using the cloud. Since 2012, this trend has been reflected in many new offerings in building automation, facility management, wireless controls and building services information management. Related trends include energy dashboards, cheap sensors, a greater awareness of the business case for energy upgrades and more government regulation and actions to cut energy use.

Megatrend #7: Cities and States Will Demand Building Performance Disclosure

Since the 2007 adoption of the Architecture 2030 standard (encouraging all existing buildings to cut energy use 50 percent and all new buildings to be "zero net energy" by 2030) and the introduction of the first "2030 District" in Seattle in 2010, group efforts to cut carbon emissions and encourage voluntary performance transparency are a major trend in the United States, capitalizing on concerns over climate change and incorporating values of openness and transparency embraced by many larger businesses and government agencies. By mid-2015, 10 US cities had functioning 2030 Districts.[10]

This trend is highlighted by more than 30 large and medium-sized US cities requiring (i.e., going beyond "encouraging") commercial building owners to disclose green building performance to tenants and buyers and, in some places, to the public.[11] This trend will spread rapidly as an easy way to monitor reductions in carbon emissions from commercial and governmental buildings and to put pressure on owners to invest in energy efficiency retrofits and renovations.

Since 2010 the European Union has mandated performance disclosure, for both new and existing buildings, under the Energy Performance in Buildings Directive (EPBD).[12] Typically, a buyer or lessee or renter gets the disclosure form during a transaction. (Figure 2.3 shows a typical example for a new building of an Energy Performance Certificate, or EPC, from the UK.) The EPC is often shown as a relative scale (similar to Energy Star) so full performance disclosure expressed as energy use intensity (EUI, recorded either in Btu/sq. ft.

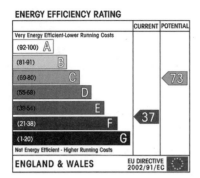
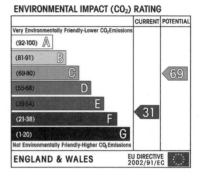

FIGURE 2.3. UK Energy Performance Certificate[14]

or kWh/m²) still lags. There are also Display Energy Certificates (DECs) for existing buildings, but their uptake is not widespread at this time.[13]

Energy Performance Certificates and Display Energy Certificates show both projected/current performance and the potential for improvement, including both energy use and carbon emissions. In Australia, disclosing a building's NABERS energy rating system became mandatory in commercial real estate transactions after 2010.[15]

Megatrend #8: Debate over Healthy Building Materials Will Become Even More Vexatious

There is little doubt that debates about healthy building products and their documentation, such as Environmental Product Declarations (EPDs), Health Product Declarations (HPDs) and various "Red Lists" of chemicals of concern that designers should avoid, will doubtless grow larger, more frequent and more contentious. The problem with EPDs and HPDs is that there are few accepted national consensus standards for determining the information that should be in an EPD or HPD, and how that information should be verified.

Nonetheless, it's easy to predict that building product manufacturers will try harder to compete for market share based on disclosure of chemicals of concern. A June 2015 report found 1,852 HPDs available from 698 brands.[16] It's also easy to foresee that industry-developed disclosure systems will compete with "verified" EPDs offered by independent rating organizations. This could lead to massive market confusion for product specifiers who must choose between proven products that they know from experience are appropriate for a given use and newer products that claim to be healthier because they meet various criteria from healthy product rating organizations, but which may not be as suitable for the intended use or as cost-effective as established products.

Megatrend #9: Solar Power Will Finally Break Through

Solar use in buildings will continue to grow, primarily because many US states will implement aggressive Renewable Portfolio Standards (RPS), while the country as a whole moves (slowly) toward zero net energy buildings. In mid-2015, 37 of 50 US states, including California, New York and Texas, had some form of RPS, mandating a specific percentage contribution from renewables to electricity supply.[17] California's RPS is the most aggressive, mandating by 2020 a one-third renewables contribution.[18]

In the United States, most electric utilities would rather build and control large central solar power plants than lose revenues to tens or even hundreds of thousands of systems feeding solar power into "their" grid. The persistence of low interest rates makes financing capital-intensive solar energy systems much easier, especially in combination with requirements in many states that electric utilities implement "net metering" programs. Such programs pay a building owner for surplus electricity produced at the same rate as power purchased from the local utility.

In the future solar power undoubtedly will be our primary electricity source, but the question is when. One expert argues convincingly that Moore's Law applies to solar power; this only makes sense, because solar modules are based on semiconductor technology.[19] So the time may come sooner than anyone expects!

Solar power has one advantage over other forms of energy efficiency (Figure 2.4): It is highly visible. Photovoltaics on the roof of a building demonstrate to employees, customers and the public that a firm or institution is committed to renewable energy and a greener future.

FIGURE 2.4. Rooftop Solar Power Is Becoming Widespread.
Credit: ©iStock.com/schmidt-z

New tools will help drive more building owners to using solar power: Google's new *Project Sunroof*, announced in August 2015, allows anyone to readily assess whether covering a rooftop with photovoltaics (PVs) would result in energy-cost savings, by combining aerial 3D models from Google Maps, historical weather data, utility prices and local financial incentives.[20] A similar system, *Mapdwell*, spun out in 2015 from an MIT research project, began offering solar prospecting services in New York and San Francisco. Mapdwell shows that each city offers multiple gigawatts of solar production potential.[21]

Solar electricity is likely to reach grid parity in the United States within the next five years, by 2020 or 2021. One estimate predicts, "If solar electricity continues its current learning rate, by the time solar capacity triples to 600 GW (by 2020 or 2021, as a rough estimate) we should see unsubsidized solar prices at about 4.5 cents/kWh for very sunny places, ranging up to 6.5 cents/kWh for more moderately sunny areas."[22]

In June 2015, Bloomberg New Energy Finance (BNEF) predicted that solar power costs would fall another 50 percent and solar investment would total $3.7 trillion in the next 25 years; of this growth,

60 percent would go to rooftop and decentralized systems and 40 percent to centralized solar power stations. BNEF also expects investment in other renewables such as wind power to total an additional $4.3 trillion.

Solar power growth is the only megatrend that is truly revolutionary, global in scope and likely to radically alter how buildings are designed, built and operated in the next ten years.

Megatrend #10: Expect a Heightened Emphasis on Water Conservation

Global awareness of the coming crisis in fresh water supply in many regions will increase as global climate change continues to affect rainfall and water supply systems worldwide. The 2012–2015 drought in California, with more than 70 percent of the state in extreme drought by summer 2015, brought water concerns to national attention in a way seldom before seen.[23]

Owing to heightened concern about future droughts' impact on water supply and cost for buildings, many building designers, owners and managers aim to further reduce water consumption in buildings by using more water-conserving fixtures, installing rainwater and graywater recovery systems, planting native and adapted vegetation instead of lawns or ornamentals, investing in more efficient cooling towers and other innovative approaches to reducing onsite water use. Case studies in my book *Dry Run: Preventing the Next Urban Water Crisis*[24] show how this is done in countries such as Germany and Australia as well as in seven major regions of the United States.

Summary

These ten megatrends will continue to drive growth of low-carbon green buildings, adoption of renewable energy in buildings and water-conserving architectural designs throughout the next ten years. We will discuss later in the book why I believe that some megatrends—a strong focus on energy efficiency, renewable energy use (mostly solar power for buildings) and water conservation—should form the core of new green building rating and certification systems.

CHAPTER 3

The Green Building Movement: A Brief History

Always do right.
This will gratify some people
and astonish the rest.

⫶ Mark Twain[1] ⫶

The green building movement started in the early 1990s with the BREEAM assessment method created by the UK government's Building Research Establishment (BRE).[2] BRE rightfully deserves recognition as the oldest, largest (in terms of project numbers) and arguably the world's most influential green building organization; its BREEAM (BRE Environmental Assessment Method) rating system influenced many other significant green building rating systems, including LEED, Green Star and Green Globes. Figure C.2 documents green building's global development over the past 25 years.

The nonprofit US Green Building Council (USGBC) formed in 1993; it developed LEED (Leadership in Energy and Environmental Design) initially as a pilot program in 1998 (LEED 1.0). The first LEED Platinum building, certified in 2001 under version 1.0, was the Philip Merrill Environmental Center in Annapolis, MD. The first technical staff member at USGBC came from BRE; he helped to create an expanded and updated standard (LEED 2.0) which debuted in June 2001, establishing the basic evaluation categories, point totals and program structure still in use today.[3] The third major iteration, LEED 2009, will be in use through June 2021.[4] LEED version 4 (LEEDv4), introduced toward the end of 2013, will become mandatory for all new projects after October 2016.

In 2001, USGBC began training individuals in the LEED system and created a professional accreditation program, the LEED Accredited Professional (LEED AP) to familiarize more people with the rating system.[5] By 2014, LEED AP and LEED Green Associate (LEED GA) numbers approached 200,000.[6] LEED's early growth was slow; however, beginning about 2005, it began to take off and grew rapidly in new office and institutional buildings, as shown in Figure 3.1. During the Great Recession, emphasis shifted to certifying existing buildings. In the past four years, new buildings have again become important users of LEED in certain market sectors.

In 2003, the Canada Green Building Council introduced LEED Canada, a LEED version adapted for Canadian conditions. In 2003, the Green Building Council of Australia (GBCA) introduced a similar standard, Green Star, for office building design. By 2014, GBCA had certified more than 800 projects. New Zealand and South Africa also use Green Star.[7] In the 2015 fiscal year, GBCA reported 253 project certifications.[8]

Later in the 1990s, BREEAM spread to Canada, where several initiatives were launched to rate green buildings and to establish criteria for sustainable design.[9] In 2000, a Canadian architect, Jiri Skopek, introduced Green Globes in Canada, as a direct offshoot from BREEAM. In 2004 he launched it in the United States, creating the first online system for collecting and assessing project information against environmental criteria. In 2005, BOMA Canada licensed Green Globes for existing buildings, calling it BOMA Go Green (now BOMA BEST). Since then BOMA Canada has certified more than 3,500 commercial buildings.[10]

The US nonprofit Green Building Initiative (GBI) licensed Green Globes from Skopek in 2006. Since then, it has become the only real US certification rival to LEED, achieving "equivalent" status in 2013 for use by the federal government.[11] In 2010, GBI established Green Globes as a national consensus standard through a process accredited by the American National Standards Institute (ANSI). In 2014, GBI launched an effort to update the 2010 standard and plans to incorporate the updated standard in 2016 in a new Green Globes for New Construction rating system.[12]

By contrast, because it grew very rapidly from 2005 to 2009, LEED became the "de facto" US green building standard during the late 2000s, supported by USGBC's 12,000 (in 2015) corporate and institutional members, and through its widespread usage by tens of thousands of architects, engineers, contractors and building owners. Unlike Green Globes, however, LEED has chosen not to submit its rating systems to a public (ANSI) consensus-based standards-setting process.

Beginning with initial efforts in these four English-speaking countries, green building rating systems have spread around the world, with more than 140 countries using the LEED system by 2014.[13] At year-end 2014, about two-thirds of LEED's international projects were in Asia, with most of those in China and India (see Chapter 4).

As of April 2015, the BREEAM scheme was used in more than 75 countries and was licensed to National Scheme Operators (NSOs) in the Netherlands, Spain, Norway, Sweden, Germany and Australia. Rating systems developed by NSOs can adapt the core BREEAM technical standard for use in their country, by demonstrating appropriate scientific support for the change, a process audited and certified by BRE.[14]

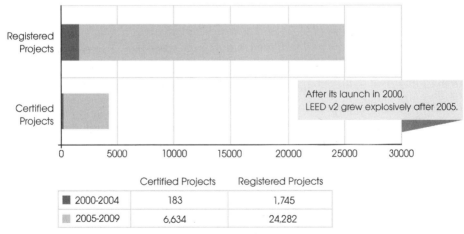

	Certified Projects	Registered Projects
■ 2000-2004	183	1,745
▨ 2005-2009	6,634	24,282

(1) Registered project totals include certified projects. Excludes projects outside US and also LEED for Homes and LEED for Neighborhood Development projects.

FIGURE 3.1. First Ten Years of LEED System Growth, 2000–2009[15]

Important country-specific rating systems are now found in widespread use in China, India, Japan, Taiwan, Singapore, Germany, France, the UAE and Qatar. However, BREEAM and LEED continue to dominate, through certifying most projects and having by far the largest organizations. In addition, more than 100 country-specific green building councils now constitute the World Green Building Council, formed in 2002.[16]

Let's take a look now at how the rating systems work, to better understand the challenges and opportunities they encounter in the marketplace. In this section, we'll examine four major systems, BREEAM, LEED, Green Globes and Living Building Challenge.

BREEAM

BREEAM is the world's foremost environmental assessment method and rating system for buildings: from 2010 through 2014, BREEAM certified 425,000 buildings; two million had registered for assessment since it first launched in 1990.[17] Of these certified buildings, 8,700 were "nondomestic" (commercial) buildings; 73 percent were in the UK; and nearly 90 percent were for new construction projects.[18] Figure C.5 shows a recent project certified at BREEAM Outstanding, the WWF Living Planet Centre, located near London.

In 2015, BRE also certified at BREEAM Outstanding (with the highest score ever at 98.4% of total points) the Edge, a large new headquarters building for Deloitte in Amsterdam, The Netherlands (Figure C.10).

Since the United States has about five times (4.9× to be precise) the UK's population, BREEAM's nonresidential certifications would be equivalent to about 33,700 US certifications, slightly more than LEED (Figure 3.2). The point here: BREEAM has had greater success in certifying commercial buildings in the UK compared to LEED in the United States, when adjusted for population.

BREEAM offers a comprehensive and widely recognized method to assess a building's environmental and social performance. It encourages designers, clients and others to think about low environ-

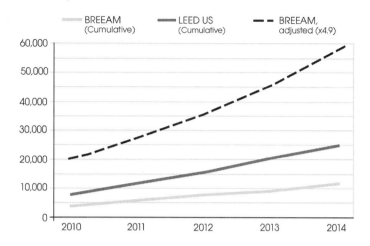

FIGURE 3.2. BREEAM vs. LEED Cumulative Project Certifications through 2015

mental impact design, minimizing a building's energy demands before considering more advanced energy efficiency and low-carbon technologies.

A BREEAM assessment is delivered by a licensed organization, using BRE-trained assessors, at various stages in a building's life cycle. As with all other green building rating programs, the assessment is designed to provide building owners, managers, developers, designers and others with:

- Market recognition for low-environmental-impact buildings
- Confidence that proven environmental practices are incorporated in the building
- Inspiration to find innovative solutions that minimize a building's impact
- A benchmark that is higher than current regulation or building codes
- A system to help reduce operating costs, and to improve working and living environments
- A standard that demonstrates progress toward corporate and organizational environmental objectives.[19]

How BREEAM Works

Each assessment looks at the following impact categories for a building's construction or operations. BREEAM rewards performance above regulation or code that delivers environmental, comfort, or health benefits. Points or credits are grouped according to environmental impact into nine categories as follows:

- Management: management policies, commissioning and site management
- Health and Wellbeing: indoor and external issues (noise, light, air, quality etc.)
- Transport: transport-related CO_2 and location-related factors
- Energy: operational energy and carbon dioxide (CO_2) emissions
- Water: consumption and efficiency
- Materials: embodied impacts of building materials, including life-cycle impacts
- Waste: construction resource efficiency and operational waste management
- Pollution: external air and water pollution
- Land Use & Ecology: site type/building footprint; a site's ecological value

Total points or credits gained in each section are multiplied by an environmental weighting that takes into account the section's relative importance. Section scores are then added together to produce a single overall score. Once the overall score for a building is calculated, it is translated into a rating level. Minimum standards (prerequisites) exist for certain categories, depending on the certification level desired.[20]

The BREEAM system contains specific rating systems to evaluate new construction, building operations, building renovations, communities and infrastructure.

Growth of BREEAM Projects

Tables 3.1 and 3.2 show project certifications from 2010 through 2014, covering nonresidential and residential projects, respectively. Nonresidential certified projects show continued growth since 2010, with

TABLE 3.1. BREEAM Certified Nonresidential Projects, 2010–2014[21]

Location	2010	2011	2012	2013	2014
UK (new)	758	1,115	1,425	1,482	1,391
UK (existing)	60	63	20	53	22
Total	818	1,178	1,445	1,535	1,413
Cumulative (since 2010)	818	1,996	3,441	4,976	6,389
Ex-UK (new)	38	78	113	248	294
Ex-UK (existing)	102	220	227	393	603
Total	140	298	340	641	897
Cumulative (since 2010)	140	438	778	1,419	2,316

TABLE 3.2. BREEAM Certified Residential Projects, 2010–2014[22]

Location	2010	2011	2012	2013	2014
UK	60,170	77,280	63,082	56,298	64,642
Ex-UK	0	4	4	2,027	10,764
Total	60,170	77,284	63,086	58,325	75,406
Cumulative (since 2010)	60,170	137,454	200,540	258,865	334,271

strong interest outside the UK in certifying existing buildings. Residential certifications show about 25 percent growth since 2010, owing mainly to the growth in projects from outside the UK.

LEED

LEED is the second largest green building rating system in the world. By the end of 2014, there were more than 30,000 certified nonresidential buildings and more than 50,000 certified residential units. Figures 3.3 and 3.4 show LEED-certified US and worldwide nonresidential projects, annually since 2010 and cumulatively over the same period. In addition to certified projects, nearly 70,000 additional projects were registered for LEED certification, mostly residential units.

Introduced as LEED v2.0 for New Construction (LEED-NC) in 2000, the LEED family began to expand with LEED for Existing Buildings (LEED-EB) in 2004.[23] LEED-EB's takeoff began when the Great Recession put a crimp on new construction during 2008–2010; by 2011 LEED-EB's cumulative certified area exceeded new construction's totals.

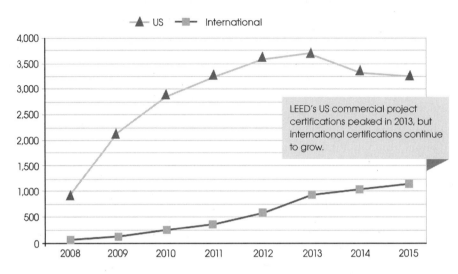

FIGURE 3.3. Nonresidential LEED Certified Projects, US and Worldwide, Annually Through 2015[24]

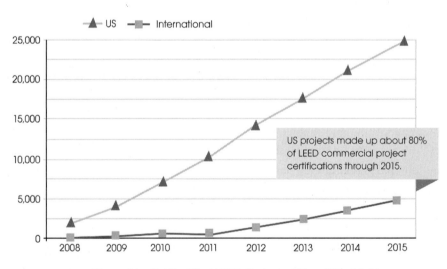

FIGURE 3.4. Nonresidential Certified LEED Projects, US and Worldwide, Cumulative Through 2015[25]

LEED certifications are issued by the USGBC-affiliated Green Business Certification Inc. (GBCI), created in 2008 (originally as the Green Building Certification Institute) to separate building certification and professional training from rating system development, which remained with USGBC.[26] GBCI also accredits professionals, either as a LEED AP, LEED AP with a particular specialty, or LEED GA.

How LEED Works

When a project desires a LEED certification, it first registers that intention with the GBCI, pays a fee and opens up a project account. Project registration is like getting engaged—it declares your intention to do something in the future, but it does not guarantee that you'll follow the process through to its conclusion and finally get married.[27] At various times during the project, project team members upload project data to LEED Online.[28] GBCI offers an end-of-construction documents certification review for new projects and a final review for all projects upon substantial completion. When a project team is finished uploading all relevant data, GBCI assigns a review team to evaluate the information; once the review is done, GBCI awards specific points earned and certifies the project at a particular level.

LEED projects are assessed in seven different categories, largely reflecting the BREEAM rating system (not surprising since LEED's developers were heavily influenced by BREEAM). In the original LEED version 2.0 rating system *for new construction* and continuing through LEED 2009 (version 3), there are seven prerequisites that must be achieved for each project to qualify for certification. In LEEDv4, prerequisites total from 11 to 14, depending on the rating system used, adding considerable cost for a project just to qualify for certification.

Originally a single system for rating new construction, when introduced as LEED version 2.0 in 2001, the LEED rating system family now encompasses 22 variations, including in LEEDv4:

- Building Design and Construction (New Construction and Major Renovations)—8 versions.

- Interior Design and Construction—3 versions.
- Building Operations and Maintenance—6 versions.
- LEED Homes—3 versions.
- LEED for Neighborhood Development—2 versions.

Once a GBCI review team awards all points achieved, they are totaled and the project is certified within one of four levels: Certified, Silver, Gold and Platinum.

Green Globes

Green Globes looks and operates similarly to both BREEAM and LEED. Green Globes began certifying nonresidential projects in 2007.[29] Contrasting with LEEDv4's 22 different rating systems, Green Globes has just four rating systems,[30] flexible enough for almost all projects. It offers an appealing approach, one that is more user-friendly, faster and less expensive. The rating systems are:

- New Construction (including major renovations).
- Sustainable Interiors.
- Existing Buildings.
- Existing Buildings—Healthcare (for facilities with more than 10 beds).

At year-end 2014, Green Globes had certified fewer than a thousand projects, giving it about a two percent market share in the United States. However, Green Globes still has merit as a rating system.

1. Green Globes for New Construction is the only US system for rating nonresidential buildings that is based on a national standard using accepted consensus procedures, "ANSI/GBI 01-2010, Green Building Assessment Protocol for Commercial Buildings." (GBI expects to release an updated standard in 2016.)
2. None of the Green Globes rating systems have prerequisites, only points. GBI maintains that if rating system categories are properly weighted, prerequisites are unnecessary to ensure meeting key sustainability objectives.
3. To secure certification, users start with an online assessment, and

then engage with accredited Green Globes Assessors throughout the process, including a final onsite review meeting, to secure a project rating, subject to final review by GBI.

4. Green Globes allows projects to subtract credit items that don't apply to a particular project from total points against which certification is assessed, thereby not penalizing projects that don't fit into a standard rating-system model.

As a result of the last three characteristics, Green Globes can often render a judgment much faster on a project's merits and offer a typical certification at a much lower total cost than a LEED project.

How Green Globes Works

A user fills out a detailed online questionnaire to start a project certification. If GBI determines that the project could achieve at least 35 percent of applicable points (the minimum certification score), GBI registers it and appoints a certified Green Globes Assessor to work with the project team, beginning with early design stages. For new construction, there is a mid-course review at the construction documents phase. Assessors make determinations about credit eligibility throughout the process. When a project is ready for a final review, the assessor conducts an onsite meeting with the project team, prepares a final report, typically 15 to 20 pages, documenting total credit points for which the project qualifies. Then, GBI issues a final certification based on the assessor's report.

Owing to the online questionnaire and the assessor's hands-on role throughout the process, Green Globes certification tends to be less expensive than LEED certification.

As an example, in 2014 a Drexel University professor published a cost analysis of the 130,000 sq. ft. Papadakis Science Building at Drexel's campus in Philadelphia, PA. Both systems certified the project at the same level (LEED Gold/Three Green Globes). The Drexel study showed that, for this project, Green Globes cost 80 percent less than LEED, after subtracting the cost for energy modeling, common to both projects. In particular, the Green Globes project did not require any consulting cost, as Green Globes' fees include most administrative costs, including professional services for the Green Globes Assessor and costs for the required onsite assessment.[31]

Why Isn't Green Globes Used More?

It's hard to see how the relative standing between LEED and Golden Globes will change in the foreseeable future, primarily because LEED has 30 times more certifications and a vastly larger "ecosystem" of accredited professionals and many more experienced consultants, designers and building owners (Figure 6.1). Most importantly, as the first and most widely used US system, for most building owners and developers, as well as the public, early on "LEED" became shorthand for "sustainability" in building design and operations.

Brands are powerful and hard to dislodge once they are established in the marketplace. However, as we shall see later on, LEED's brand is strong only in relatively few market segments.

Living Building Challenge (LBC)

The Living Building Challenge debuted in 2006 as "LEED on Whole-Grain Natural Steroids,"[32] offering only prerequisites and no points, and with the perspective that a building ought to be regenerative, i.e., positively good and not just "less bad." Sounds good, but as usual, "The devil is in the details, and it's all details."

How the Living Building Challenge Works

LBC posits 20 "imperatives" (prerequisites) grouped into seven different categories called "petals" (think of a flower) that a building must have to be judged "living" by this standard. The petals comprise:

1. Place.
2. Water.
3. Energy.
4. Health and Happiness.
5. Materials.
6. Equity.
7. Beauty.

Five petals correspond more or less to typical categories in other established green building rating systems, but then LBC goes beyond just physical performance to impose value judgments about what constitutes a "living" building and introduces new categories, "equity" and "beauty," to implement its philosophical agenda.

One unique feature that I like: For zero net energy use, before it will award certification, LBC insists on having at least one full year's operating data, proving that the building achieved in practice the projected result.[33] This is similar to the assertion in my recent book, *The World's Greenest Buildings*, that "only performance matters."[34] Far too many LEED-certified buildings have been shown to use more energy in practice than projected by design energy models (in some cases, twice as much) to be sanguine about energy modeling accurately reflecting future performance.[35]

Growth of LBC Projects

However, in nearly nine years since its introduction in late 2006, by August 2015 LBC had awarded "Living Building" certification to just nine buildings,[36] a strikingly poor record for a system designed to be a major change agent for buildings.[37] Excepting the 45,000 sq. ft. (net leasable area) Bullitt Center in Seattle, WA, (Figure C.7) and the 24,000 sq. ft. Center for Sustainable Landscapes at Phipps Conservatory[38] in Pittsburgh, PA, the remaining seven buildings certified as "Living" are either too small or too specialized to serve as practical models for commercial buildings.

LBC also awards a Petal certification to a building that meets at least one "top three" Petal (zero net energy, zero net water, or materials selection), demonstrates compliance with the requirements of two other Petals, and also meets two individual imperatives called "Limits to Growth" and "Inspiration and Education."[39] While perhaps encouraging and recognizing some meritorious projects, Petal certification is a far cry from LBC's advertised *all or nothing* approach to green building.

Additionally, LBC offers Zero Net Energy certification that "recognizes building projects that achieve the Energy Petal, along with a subset of Imperatives within Place and Beauty Petals."[40]

These two other certifications have also met with scant market interest: by August 2015, LBC had awarded Zero Net Energy certification to some 11 buildings (four residential) and a "Petal" certification to six projects (three residential).[41]

Perhaps it's better to think about the Living Building Challenge as a laboratory prototype, or even as a think tank, rather than as a pragmatic green building certification program. LBC's orientation is toward "pushing the envelope" of radical building design instead of merely seeking to influence current commercial design practice for individual buildings.

National Green Building Standard (NGBS)

NGBS is a rating system similar to LEED for Homes. Discussed in more detail in Chapter 5, NGBS certifies single-family homes and multifamily structures/developments to a consensus standard called ICC 700, the only residential green building rating system approved as an American National Standard by ANSI.[42] As of January 2016, NGBS had certified 69,819 homes,[43] roughly equivalent to LEED for Homes, which announced its 50,000th certification in January 2014 and which had certified 70,000 US residential units through April 2015. LEED tends to dominate the multifamily category, with about 70 percent of its units coming from that source, while NGBS and regional/local rating systems tend to dominate the single-family category.[44]

By some estimates, other local or regional home rating systems constitute the market's other 50 percent, giving a *total US certified* green home market size in mid-2015 of about 220,000 certified homes and multifamily units, a miniscule fraction of 115 million housing units.[45]

NGBS is unlikely to grow rapidly nationally, given the competition from LEED and from other localized systems documented in my book *Choosing Green*.[46] This local competition will continue to control about half the US new-home rating market, but their localized nature makes it difficult for any to scale up enough to expand the overall market. However, as we'll see in Chapters 5 and 8, there may be a growing market for LEED certification in multifamily homes because of its streamlined delivery model.

Comparing Three Leading US Green Building Rating Systems

In 2012, in a study of green building certification systems for the US General Services Administration (GSA), Pacific Northwest National Laboratory (PNNL) examined three systems: LEED, Green Globes and LBC. Their study concluded that for new construction, Green Globes met 25 federal requirements for sustainability; LEED met 20 requirements; and the LBC only 15. In other words, LBC (version 2.0) was not deemed usable for meeting sustainability goals for the US's largest property owner and operator, the federal government. The study concluded that either LEED or Green Globes would meet federal requirements for high-performance, sustainable buildings.[47] The study rated LEED as marginally superior to Green Globes for evaluating existing buildings.

If life were fair and one only considered such factors as cost, speed and ease of use, then Green Globes would be preferred to LEED. However, life isn't fair: LEED had a 10-year head start, LEED has built an extensive ecosystem with advocates and providers, and LEED created a strong continuing education system, so it has gained the lion's share. However, as we'll see later in the book, the biggest competitor for both LEED and Green Globes is the "none of the above" or "do-it-yourself" approach, owners who choose not certify under either system, but nonetheless maintain that they are building or operating a "green" building.

Summary

Green building rating systems have been around for 25 years and seem to have settled into their respective niches in the new and existing buildings market. Currently, they all assess similar issues; the only real difference is the delivery model. In the United States, LEED and Green Globes compete for market share, with LEED the dominant system. In Canada, LEED Canada and BOMA BEST (which only certifies existing buildings) compete.

In the UK and in some other parts of western Europe, BREEAM is the dominant rating system. For example, Figure C.17 shows a project in Belgium, a new headquarters for the Flanders Red Cross, certified at the BREEAM Outstanding level in 2014.

In other countries, there are many country-specific systems that compete for project certification revenues, along with LEED's and BREEAM's international versions. The entire green building market is ripe for a disruptive innovation in product features and delivery models, because outside of Singapore, none of these systems commands a significant share of the overall buildings market.

CHAPTER 4

The US Green Building Movement Today

*Many years ago there was an Emperor so exceedingly fond
of new clothes that he spent all his money on being well dressed....
"But he hasn't got anything on!" the whole town cried out at last.
The Emperor shivered, for he suspected they were right.
But he thought, "This procession has got to go on."[1]*

Since 1990 green building certification has spread from just an idea promoted by a few people to a global phenomenon. In this chapter, we dive deeper into the current situation and provide a review of the six-year period, 2010 through 2015. This period is long enough to establish some trends that are likely to persist into the following five years. We conclude that green building certification today works for only a limited portion of the overall buildings market.

We won't deal further in this chapter with Green Globes or Living Building Challenge, as LEED certifies about 95 to 98 percent of all projects and, at this point, green building certification's fate in the United States will be determined largely by LEED's fate. Unfortunately, that is not looking so good at the present time.

LEED is failing to grow in the United States, its largest and most important market! At year-end 2015, LEED had certified about 0.7 percent of US nonresidential buildings and less than four percent of total nonresidential building area.[2] For residential units, LEED certifications amount to less than 0.04 percent of the 115 million US residential units (and less than two percent of the new single-family housing units built just in the 2010–2015 period).

If cutting carbon emissions is our overriding concern and if green building should set its sights on the target of transforming at least 25 percent of buildings, *there is no way that LEED will ever reach this goal.* Using its own definition of this target and after 15 years of effort by many thousands of professionals, isn't it time to admit that LEED is basically an elite rating system, used primarily by wealthy building owners, which will never reach the mainstream?

USGBC's founder and first CEO, David Gottfried, makes the very clear point that "numbers and data matter," both those that "prove the case for green building and those that challenge the success we've had to date."[3] He further observes that "We've forgotten to involve all buildings in the green building game...what about the millions of other buildings, most of them small? We need them to be part of the green building improvement game."[4] We'll return to this point often.

LEED Data Sources

Our information is based on data extracted and analyzed primarily from USGBC's LEED Project Directory,[5] a public database, which we have carefully reviewed. We stopped the original analysis at the end of 2014, since data indicated that LEED project registrations and certifications had stopped growing, both in the United States and internationally. Further analysis of LEED project data through the end of 2015, using the LEED monthly updates,[6] supported that choice. You can find LEED project data for 2010 through 2015 in the appendix, "2015 LEED Projects Update."

In addition to not reaching takeoff velocity for US building certifications, LEED registration and certification growth leveled off after 2010 into a more steady state, mostly involving large buildings, despite building construction's 36 percent growth during that period.[7] Figure 4.1 shows growth in US LEED registrations and certifications since 2005 for nonresidential projects. Project registrations represent the best indicator of current LEED project activity because they show projects that *started* in a given year with the intention to certify once they were built. One can easily see that growth has tailed off and that US LEED registrations for nonresidential projects have plateaued at fewer than 4,500 projects per year.

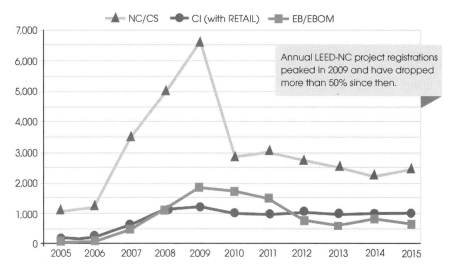

FIGURE 4.1. Annual LEED Nonresidential US Projects, 2005–2015

This is a classic chart of market saturation, typically signaling a need to introduce new products and new approaches (and not just to update old products) to rekindle market growth. As we will show in later chapters, USGBC and LEED appear to have exhausted their good ideas and are relying on momentum, along with a small army of LEED advocates, to continue US market activity levels, while turning their attention for future growth to the international market.

More important than LEED's past or current growth is the task ahead. Figure 4.2 indicates that through the end of 2015, LEED had certified less than four percent of the total area of commercial buildings in the United States. While it has certified less than one percent of buildings, the percentage of commercial area may be a better measure of success and market penetration, since a LEED project's average size, more than 110,000 sq. ft., is significantly larger than the US commercial building stock's median size of about 5,000 sq. ft. (Table 7.1).

Considering that LEED adds only about 0.5 percent of US building area (about 417 million sq. ft. of certified project area) and only about 0.05 percent of US buildings (about 2,750 buildings) to its certification totals each year, it's clear that LEED will never reach its stated goal: 25 percent of US buildings certified to higher-level green building standards.

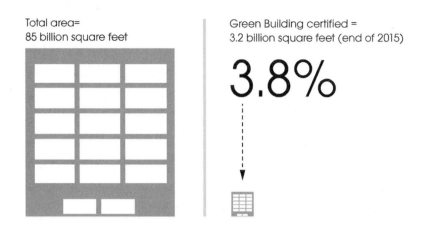

Total area=
85 billion square feet

Green Building certified =
3.2 billion square feet (end of 2015)

3.8%

FIGURE 4.2. US Building Area vs. LEED Certified Building Area, End of 2015[8]

One might never conclude from the constant stream of publicity and promotional material from USGBC that there's "trouble in paradise," but the numbers show otherwise. There's another way to gauge what's happening: follow the money.

As shown in Figure 4.3, USGBC's operating revenues have been declining since the high point in 2009, indicating diminishing support not only for LEED but for the organization as well.[9] The revenue stream for the Green Business Certification Institute, USGBC's LEED operating unit, is below 2009 levels, but not as sharply. USGBC's stated membership—about 12,200 companies in 2014—is down almost 40 percent from its pre-recession high of nearly 20,000 corporate members. In its 2014 annual report, USGBC reported a consolidated (with GBCI) operating loss of more than $6.5 million, on top of a $5.6 million loss in 2013, so it's clear that the adverse profit trend of the previous five years is continuing.[10]

My intent in this book is not to denigrate what USGBC and LEED have accomplished, more importantly what its thousands of volunteers (including myself) have accomplished over the past 15 years, but to indicate forcefully that *the era of LEED's rapid growth is long past and it's time for a new approach, time for a real evolution (revolution) in how we approach green building, not just an incremental change in a system that isn't working.* One would think that a nonprofit like USGBC

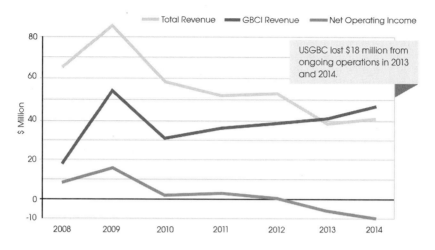

FIGURE 4.3. USGBC and GBCI Revenues and Operating Profits, 2008–2014[11]

whose stated mission is "transforming the building industry within one generation" would be the first to step up and say that its current approach is not working, but that has not been the case.

In the next chapter, we'll discuss why green building is vitally important to our low-carbon future, and in Chapter 6 we'll discuss in greater detail what USGBC and LEED have done right, followed in Chapter 7 by our analysis of what they've done wrong, but here I want first to convince you that there is a problem and second, that trends are moving in the wrong direction.

More importantly, I would like this book to convince USGBC to change course, primarily by changing LEED's structure, what it contains and how it goes to market. The most important message is that the marketplace's judgment, while harsh, is right: LEED's current system is too costly and creates too little value to appeal to most building owners and facility managers.

LEED in the United States since 2005

Let's look at some LEED programs in detail. In general, there are three basic programs for the nonresidential ("Commercial") building sector and three versions for the residential sector.[12] In the commercial building sector, the LEED programs are:

1. Building Design and Construction: encompassing LEED for New Construction and LEED for Core & Shell, including all forms of new construction and major renovations, such as for retail, healthcare, schools, data centers, etc.
2. Interior Design and Construction: primarily tenant improvements in office and retail fit-outs (think of Starbucks stores).
3. Building Operations and Maintenance: evaluations of ongoing building and facility operations.

In the residential sector, the rating systems are LEED for single-family (detached) residences, multifamily low-rise (up to three stories) and multifamily mid-rise (four to eight stories). Some high-rise residential buildings may choose to certify under the LEED for New Construction (and not the Homes) category, particularly those that are mixed-use projects such as office/residential, hotel/residential, etc.

Figure 4.4 shows program growth from 2005 (the takeoff date for LEED in the United States) through 2015. The data show new project registrations, which are still the best market measure for interest in LEED because certifications, especially for new buildings, can lag anywhere from two to four years after registration, as a building is constructed, occupied and eventually certified.

The figure shows the three major components of LEED commercial project registrations: new construction (NC), including core and shell offices (CS); commercial interiors or remodels (CI); and existing building operations and maintenance (EB/EBOM). From a high point in 2009, LEED project registrations for new buildings in the United States have declined 60 percent to 2015 levels of less than 2,600. Some of this decline owes to the Great Recession, but for new building construction the recession basically ended in 2011.

During the past six years, LEED-certified US nonresidential project area has tapered off by about six percent, as shown in Figure 4.5. Along with the reduction in new project registrations, this trend presages a continued decline in LEED project certifications in the years ahead. In 2015, the number of LEED US nonresidential registered

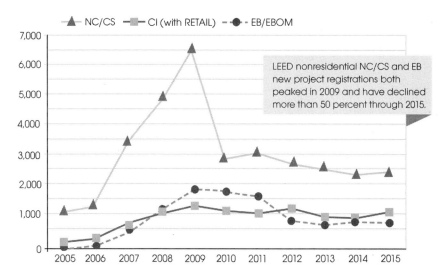

FIGURE 4.4. LEED Registered Nonresidential US Projects, by Rating System, 2005–2015

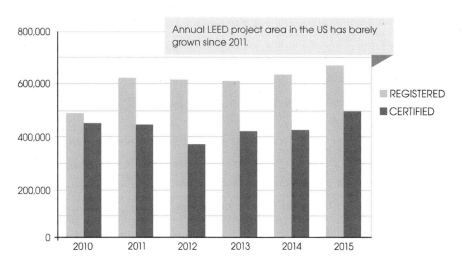

FIGURE 4.5. LEED Nonresidential US Projects, by Area 2010–2015

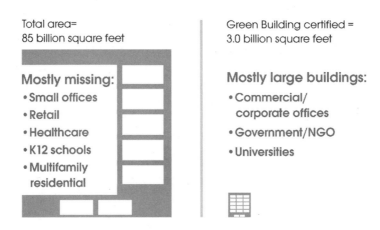

Total area=
85 billion square feet

Green Building certified =
3.0 billion square feet

Mostly missing:
• Small offices
• Retail
• Healthcare
• K12 schools
• Multifamily
  residential

Mostly large buildings:
• Commercial/
  corporate offices
• Government/NGO
• Universities

FIGURE 4.6. What's Missing from US LEED Projects?

projects increased about eight percent from 2014 levels, but certified project numbers stayed the same.

There's always more to the story than just overall numbers, and in Chapter 8 we will look at specific rating systems and specific building types for further guidance, but Figure 4.6 indicates what's missing: a large part of the US commercial and institutional building industry.

LEED Projects in 2014

We analyzed LEED project data in detail for 2014. For both new project registrations and project certifications, Table 4.1 shows numbers and total area, indicating the current size of the US and international markets for LEED projects. Figure 4.7 shows the six-year trend, indicating that LEED certified projects are in steady state, with no real growth during the past five years.

The US Construction Market

The broader US construction market provides the backdrop for evaluating LEED's success in penetrating various market segments. Table 4.2 shows construction put in place in 2014, by dollar value.[13] Clearly, the United States is a huge market for all new construction projects, but how many are LEED projects?

TABLE 4.1. Nonresidential LEED Projects, US and International, 2014

Certification System	Registered Projects	Total Area (MM sq. ft.)	Certified Projects	Total Area (MM sq. ft.)
New Construction/ Core & Shell	1,109	184	1,016	136
Existing Building O+M	784	318	545	185
Commercial Interiors	620	30	844	26
Healthcare*	114	16	81	10
Higher Education*	191	18	190	15
K12 School*	203	19	201	20
Office*	389	34	284	21
Retail**	583	6	196	4
US Total	**3,993**	**625**	**3,357**	**417**
International	**2,022**	**625**	**1,048**	**240**
Global Total	**6,015**	**1,250**	**4,405**	**657**

* New Construction only (NC/CS)
** Both New Construction and Interior Fit-outs, 50% to each

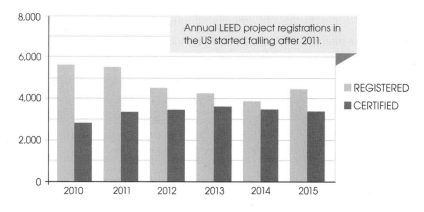

FIGURE 4.7. LEED Commercial Projects in the US, 2010–2015

Assuming a $150/sq. ft. construction cost, the commercial new construction market in 2014 totaled about 1.77 billion sq. ft. ($265 billion/$150/sq. ft.) From Table 4.1, LEED registered NC/CS projects totaled 274 million sq. ft., about 15 percent. That is no mean feat, but it still means that LEED is falling farther behind *each year*, in terms of certifying total commercial building area (new and existing), since

TABLE 4.2. US Construction Market Size, 2014 ($ Billion)

Category	Private	Public Sector	Total
Office	36.9	7.7	44.6
Lodging	15.6	0.5	16.1
Commercial	55.5	1.8	57.3
Healthcare	28.9	10.1	39.0
Educational	16.4	62.0	78.4
Religious	3.6	0.0	3.6
Public Safety	0.2	9.1	9.3
Amusement	7.8	8.9	16.7
Total Nonresidential	**164.9**	**100.1**	**265.0**
Residential	349.0	5.2	354.2
Total	**513.9**	**105.3**	**619.2**

85 percent of all new buildings each year are NOT registered. In existing buildings, LEED registered about 318 million sq. ft. in 2014, representing about 0.4 percent of existing buildings' 85 billion sq. ft.[14]

Taken together, LEED registered about 0.7 percent of US building area in 2014. What this means, effectively, is that LEED certifications by 2020 may never represent more than seven percent of commercial building area and not even one percent of the number of commercial buildings. *In this respect, green building in the United States has definitely hit the wall and something new is needed.* Over time, LEED will never get even close to USGBC's originally stated goal for market transformation: certifying 25 percent of all US buildings.

International

USGBC and GBCI have done considerable work since the mid-2000s to create an international version that can be used by projects worldwide, with some initial success. As of mid-2015, 140 countries used LEED. For the top ten countries using LEED internationally, projects in just three counties, Canada, China and India, constituted 70 percent of total certified project area.[15]

Nonresidential. International projects represented about one-third of all new LEED registered projects in 2015 and about 20 percent of all

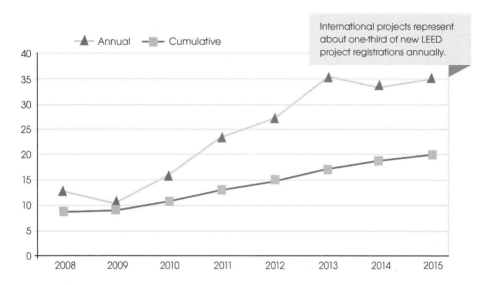

FIGURE 4.8. International LEED Projects, Percent of Totals, 2008–2015

projects to date (Figure 4.8). That's fantastic, of course, but the most likely scenario is that LEED is skimming the cream in various countries and that longer-term, project registration trajectories will likely follow US trends. Figure C.3 shows a typical example of a LEED international project, a new high-rise office in Istanbul, Turkey, for an international financial company.

In 2014 and 2015, new LEED international project registrations (including residential projects) showed about a 30 percent drop from 2013 levels (Figure 4.9). An analysis of 2015 results shows an increase in registered projects from 2014 levels but still below 2013 totals (see the appendix).

There may be three reasons for international project growth to slow, even as those projects at year-end 2015 comprised about 20 percent of all LEED-registered projects.

1. There are many national rating systems that compete with LEED, including those in major construction markets such as India and China, the Middle East, Europe and Canada.
2. Some countries will seldom use LEED's international certification program, because their own systems are well entrenched. These

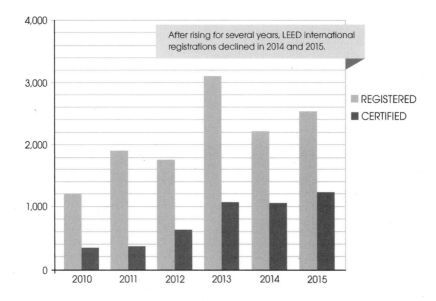

FIGURE 4.9. Growth of LEED International Projects, 2010–2015 (Including Residential)

countries include Germany, most of the Middle East, Australia, the UK, France, New Zealand, Japan, South Africa and Singapore. In those countries, LEED's main use comes from global real estate organizations that only want to use one system and also from large developers accustomed to using LEED.

3. There are other international rating systems that are quite good and offered by well-funded nonprofit organizations, primarily the BREEAM system which dominates the United Kingdom and has a strong following in Western Europe.

Residential. As of June 2015, LEED for Homes outside the United States had certified 221 residential units and had registered an additional 769 units, for a total of just under one thousand homes. Ninety percent of *certified* projects were in Saudi Arabia and 88 percent of *registered* projects were in seven countries: China, United Arab Emirates, Turkey, Montenegro, Haiti, Cayman Islands and Guam.[16] (Note that because of multifamily projects, there will always be more units registered and certified than "projects.") It's fairly clear from these data that LEED for Homes has attracted very little interest internationally.

Summary

The American novelist F. Scott Fitzgerald wrote, "The test of a first-rate intelligence is the ability to hold two opposed ideas in mind at the same time and still retain the ability to function."[17] Since I respect your intelligence, I would like to give you the same opportunity! Considering LEED, you might hold one (or both) of two opinions.

LEED is a colossal success: Starting from just an idea and a few committed people 20 years ago, USGBC and LEED now sit astride the world green building market with the LEED 2009 certification system used in 140 countries and with $40 million per year in revenues flowing through the GBCI (renamed Green Business Certification, Inc. this year). LEED has become synonymous with green building worldwide, and USGBC is primarily responsible for this success story.

LEED is a colossal failure: Given USGBC's self-proclaimed mission to transform the building industry, it has become instead a rating system used primarily by "the 1 percent": high-end office developers and owners, large global corporations, and some top US colleges and universities. However, entire large sectors of the buildings industry— smaller offices, retail, healthcare, K12 education, etc.—ignore LEED entirely or use it sparingly.

According to USGBC founder David Gottfried, writing in 2014,

> The LEED green building rating system only certifies buildings that are determined to be at the leadership level (the "L" in LEED).... Initially, USGBC decided that the minimum bar for the "L" level for LEED certification would begin with the top 25 percent of buildings (LEED Certified-level), and then go up from there to higher levels of sustainable performance.[18] (See Figure 4.10)

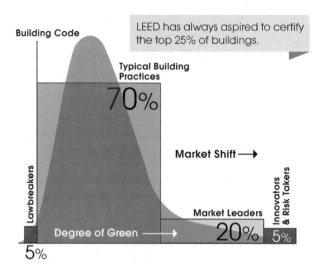

FIGURE 4.10. LEED Originally Aspired to Certify the Top 25 Percent of Buildings to Induce Market Transformation

As Pamela Lippe writes in the Foreword, "Getting a smaller and smaller segment of the population to jump higher and higher is not going to get us where we need to go." If the goal all along was to get 25 percent of buildings LEED-certified and we're *not even at one percent* after fifteen years of hard work by tens of thousands of highly skilled and motivated professionals, wouldn't you call that a failure?

Take your pick; either interpretation is possible—the choice is yours! My thesis here is straightforward: If reducing total global carbon emissions is the overriding environmental, economic and public health issue of our time, LEED (alone) is not going to get the job done, not today, not tomorrow, *not ever*. And that means it's past time to look for an alternative approach or approaches—and fast!

CHAPTER 5

The (Business) Case for Green Building

Sustainable development is a fundamental break
that's going to reshuffle the entire deck.
There are companies today that are going to dominate
in the future simply because they understand that.

Francois-Henri Pinault[1]

If green building certification has indeed hit the wall, can we still say that green buildings are *necessary*? Many surveys indicate that green building measures such as energy efficient design, water conservation, use of recycled-content materials, better indoor air quality, daylighting and local habitat preservation have become more widespread.[2] However, these same surveys show that LEED projects may not achieve significant performance gains, with the average energy efficiency *improvement* in LEED buildings in Canada, for example, only 9 percent better than conventional construction.[3]

Carbon Reduction in Green Building

Here's the central issue: Without a strong and focused green building certification review, most people will be tempted to do "green" halfheartedly and will not achieve carbon emission reductions vital to mitigating climate change.

Where did this concern with buildings' climate-change effects originate? It began with a seminal article in 2003 by New Mexico architect Edward Mazria, in which Mazria recast conventional energy use statistics to focus on end-uses (rather than sources such as electricity, gas, etc. or sectors such as industry, residential, etc.) and found that nearly 40 percent of energy end-use in the United States and other developed countries came from homes and commercial/industrial buildings, about half from each source.[4] With that article the green building industry had a reference point, a rationale for improvements in building energy efficiency, one that dealt directly with the source of global climate change.

Added to energy use for building operations is embodied energy in building materials used for construction and renovation. When these are included, buildings in the developed world contribute nearly 50 percent of total energy use and carbon emissions. Add to that the transportation energy use required for servicing buildings and people traveling to and from work, and it's clear that it is critical for green building to focus on energy and location if we are to deal with carbon emissions.

If reducing carbon emissions is our central, critical problem, an effort that will take decades, then why must a green building include *all* of the other environmental categories embraced in leading rating systems?

LEED has problems gaining marketplace acceptance, as we showed in Chapter 4, largely because it has become a Christmas tree, loaded up with ornaments showcasing most conceivable environmental and sustainability concerns.

Having too much stuff in a rating system leads to 700-page reference manuals, arcane rulings to cover every conceivable permutation in design and operations, and a review system that causes never-ending headaches for consultants and stretches out the review process for many months, even years, causing owners to question what the review accomplishes.

A LEED consultant with more than 15 years' experience recently told me of a classic example: A campus has a strict "tobacco-free" policy, prohibiting not only smoking but also tobacco chewing and e-cigarettes. LEED requires, as a prerequisite, that each building be smoking-free with specific signage to that effect; in this case, reviewers refused to take into account the campus' tobacco-free nature and insisted that individual buildings display LEED-approved *No Smoking* signs before they would approve this particular prerequisite for the project!

It's time to refocus our green building thinking around carbon and things that drive carbon emissions from buildings: gas, diesel and electricity use; transportation; water use (considering the water/ energy nexus); waste generation; and purchasing (reflecting embodied carbon in materials).

This is where LEED has stalled. The rating system for existing buildings, LEED-EBOM, may be considered fundamentally pointless by 70 percent of potential users, as it only accepts projects that are already in the top 31 percent of energy use (an Energy Star score of 69 or higher is a prerequisite).[5] LEED-EBOM certified fewer than 550 projects in 2014 and its market acceptance, measured by new project registrations, fell by more than 50 percent from a high point in 2009, as shown in Figure 8.8.

The Business Case for Commercial Projects

Yet there must be reasons why so many US projects still pursue certification via LEED and Green Globes and why so many international projects are pursuing LEED and BREEAM. Putting aside particular issues with LEED's prerequisites, which we'll take up in subsequent chapters, *there is still a strong business case for green building practices and certification*, whether in new construction, major renovations, interior remodeling or existing building operations, and no matter what the ownership: public, private, nonprofit, corporate, education, healthcare, retail, etc.

The business case involves four key benefits, as identified to me by a major financial corporation's facilities director who uses both LEED and Green Globes to certify projects:

1. Defining what's "sustainable": providing design teams with a *menu of options* for green design and providing project teams with a clear definition for sustainability.
2. Saving future operating costs, especially for energy (and to some degree water).

3. Providing credibility for reporting to stakeholders what you're doing.
4. Responding to government incentives/requirements, where they exist.

Let's examine these benefits individually.

Defining What's Sustainable

Green building certification's core value, especially for new buildings, is to help define what's sustainable and to give design and construction teams a *menu of options*. It's like a restaurant menu: You get one appetizer, one soup, one salad, one entrée and one dessert; and you have a complete (green) meal. With green buildings, we want to reduce direct and indirect carbon emissions, conserve water, reduce resource use and encourage healthy indoor environments, so we can have a more ecologically sound built environment (Figure 5.1).

One expert, Curtis Slife, puts it this way:

> All certification systems make an attempt to develop effective communication for owners and for facility managers. They explain what sustainability means and how you achieve it. They bring clarity to the language of what we're trying achieve. Obviously, if we can achieve more sustainable structures, it benefits the whole community, but I think rating systems provide a structure for the application of sustainable processes and practices. They provide reward mechanisms for investment in sustainable design and practices. When we didn't have these programs, sustainability was difficult to sell. Everybody had what they thought was a best practice, but frankly it wasn't.[6]

Today's problem is that *the menu is way too long*. For LEED or Green Globes the choices read like those in the 30-page menu at The Cheesecake Factory restaurant chain: by the time you finish reading it, you're no longer hungry!

FIGURE C.1. Ten Green Building Megatrends

1990
BREEAM/Austin Green Builder launched for new offices

1992
US ENERGY STAR (PCs and Monitors) launched; **UN Rio Conference**

1993
US Green Building Council (GBC) formed

2004
LEED for **Existing Buildings** introduced; **Green Building Initiative** formed

2003
First **Green Star** rating tool, **GBC Australia** formed

2002
First **Greenbuild** (US) conference/ **Canada GBC** & **World GBC** formed

2005
Singapore Green Mark/ LEED 2.2/ BOMA Canada launches **BOMA BEST**

GREEN BUILDING
The First **25** Years

2007
German Sustainable Building Council(DGNB) formed; 1st US **Green Globes** certifications/ **Architecture 2030 targets** endorsed

2006
Living Building Challenge launched as "LEED on Steroids"

2016
LEEDv4 Mandatory/ **Green Globes ANSI Standard** updated

2015
World GBC in 100 countries

2014
LEED used in more than **140 countries**/ **BREEAM** updated/ **LBC v3.0**

FIGURE C.2. Green Building Timeline

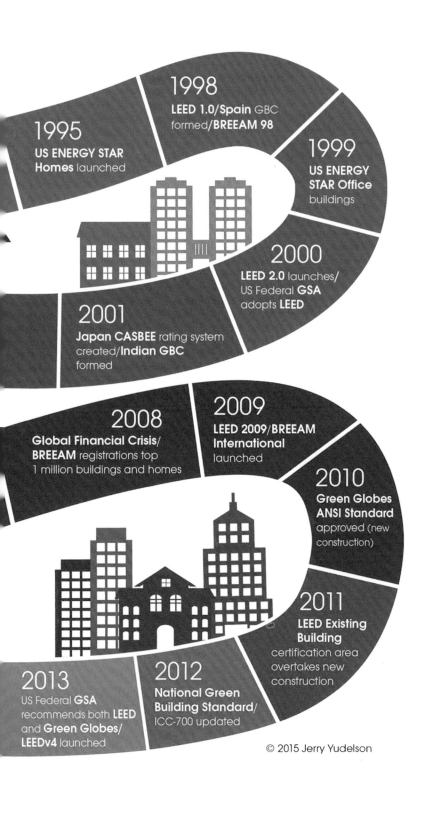

1995
US ENERGY STAR
Homes launched

1998
LEED 1.0/Spain GBC
formed/**BREEAM 98**

1999
**US ENERGY
STAR Office**
buildings

2000
LEED 2.0 launches/
US Federal **GSA**
adopts **LEED**

2001
Japan CASBEE rating system
created/**Indian GBC**
formed

2008
**Global Financial Crisis/
BREEAM** registrations top
1 million buildings and homes

2009
**LEED 2009/BREEAM
International**
launched

2010
**Green Globes
ANSI Standard**
approved (new
construction)

2011
**LEED Existing
Building**
certification area
overtakes new
construction

2013
US Federal **GSA**
recommends both **LEED**
and **Green Globes/
LEEDv4** launched

2012
**National Green
Building Standard/**
ICC-700 updated

© 2015 Jerry Yudelson

FIGURE C.3. Allianz Tower, Istanbul, Turkey. Designed by FXFOWLE for Renaissance Development and completed in 2014, the Allianz Tower rises 40 stories and contains 930,000 gross sq.ft. (83,000 gross m²) of leasable space. Certified as the first LEED Platinum high-rise in Turkey, the project projects a 26 percent energy-use reduction compared with a standard building. Allianz Tower typifies LEED international projects: a major commercial office building for a large corporate client. Credit: Photography by David Sundberg/Esto. Courtesy of FXFOWLE.

FIGURE C.4. Federal Center South, Seattle, WA. Designed by ZGF Architects LLP and completed in 2012 on an overall $74 million project budget, this adaptive reuse achieved LEED Gold certification. Federal Center South Building 1202 transformed a 4.6-acre brownfield site into a highly flexible and sustainable 209,000-sq. ft. regional headquarters for the U.S. Army Corps of Engineers (USACE) Northwest District. Under a performance-based contract, GSA retained 0.5% of the contract value until it verified energy performance one year after occupancy. Credit: ZFG Architects LLP. Photography by Benjamin Benschneider

FIGURE C.5. WWF Living Planet Centre, Woking, UK. Certified at the BREEAM Outstanding (highest) level and designed by Jason Bruges Studio, the solar-powered complex serves as a home base for 300 employees. Credit: www. aireyspaces.com

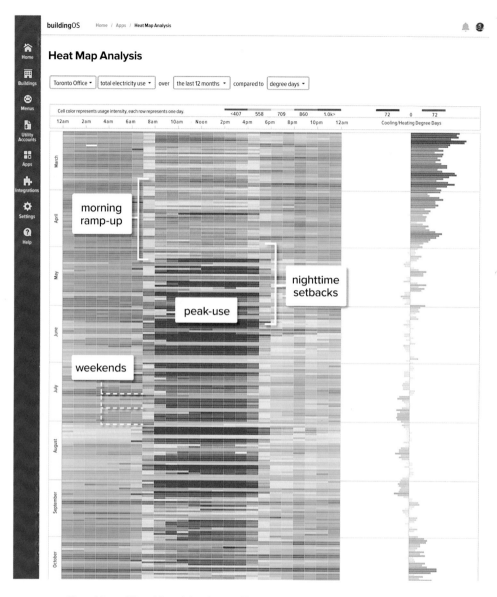

FIGURE C.6. Heat Maps Allow Visual Analysis of Energy Use Patterns. Credit: Lucid

FIGURE C.7. The Bullitt Center, Seattle, WA. A certified "Living Building," The Bullitt Center is a six-story commercial office building in Seattle. Completed in 2013 for a total project cost of $32.5 million, it was certified in 2015 also as a LEED Platinum building. It has a verified zero net energy performance, and in 2014, its first year of operation, it produced 60 percent more energy than it used. Credit: ©Nic Lehoux

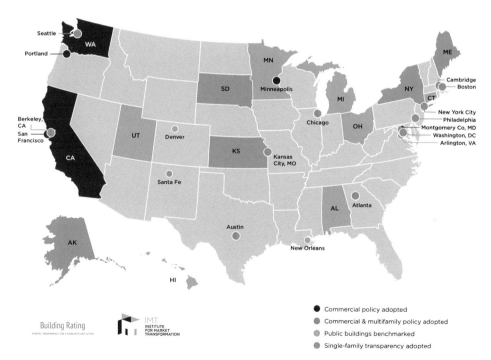

FIGURE C.8. US Building Benchmarking and Transparency Policies, June 2015.
Credit: © The Institute for Market Transformation, http://www.imt.org, July 2015

FIGURE C.9. J. Craig Venter Institute, La Jolla, CA. Certified at LEED Platinum and designed by ZGF Architects LLP, the J. Craig Venter Institute in La Jolla, CA, is a zero net energy research laboratory. This highly efficient building uses a 500 kW rooftop solar PV array to provide renewable energy. Credit: Nick Merrick © Hedrich Blessing

FIGURE C.10. Certified at BREEAM Outstanding (with the highest score ever at 98.4% of total points), the Edge is a new headquarters building for Deloitte in Amsterdam, The Netherlands. This 40,000 m² (430,000 sq. ft.) office building, designed by PLP Architecture of London for the Dutch developer OVG, also claims to be the "smartest" building in the world, with more than 28,000 sensors delivering real-time data to the building's operator. Credit: © Ronald Tilleman

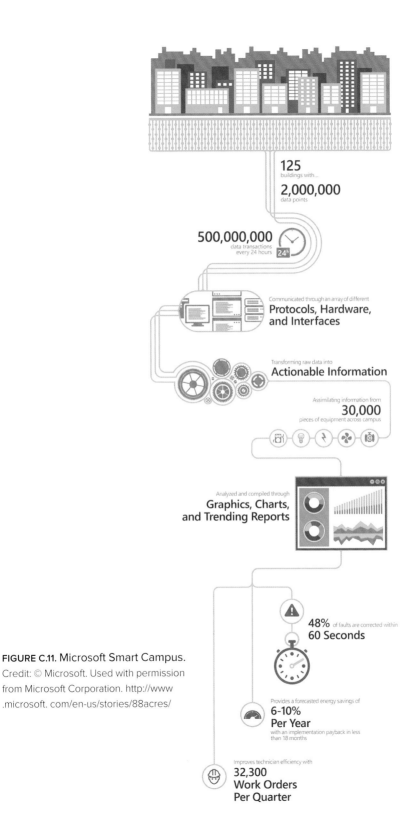

125 buildings with...

2,000,000 data points

500,000,000 data transactions every 24 hours

Communicated through an array of different
Protocols, Hardware, and Interfaces

Transforming raw data into
Actionable Information

Assimilating information from
30,000 pieces of equipment across campus

Analyzed and compiled through
Graphics, Charts, and Trending Reports

48% of faults are corrected within
60 Seconds

Provides a forecasted energy savings of
6-10% Per Year with an implementation payback in less than 18 months

Improves technician efficiency with
32,300 Work Orders Per Quarter

FIGURE C.11. Microsoft Smart Campus.
Credit: © Microsoft. Used with permission from Microsoft Corporation. http://www .microsoft. com/en-us/stories/88acres/

FIGURE C.12. Energy Dashboard Display on PC or Tablet. Credit: Switch Automation

FIGURE C.13. Lucid's BuildingOS Analytics. Credit: Lucid

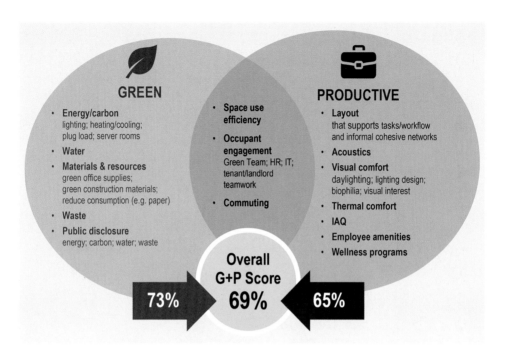

FIGURE C.14. Green + Productive Workplace Issues. Credit: Simone Skopek and Bob Best

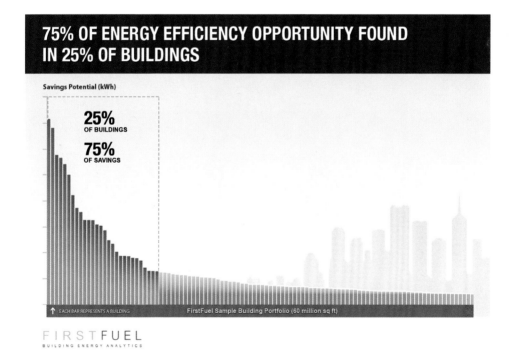

FIGURE C.15. Energy Efficiency Opportunities in Commercial Buildings. Credit: © FirstFuel Software

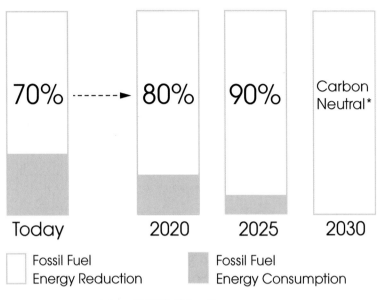

70% ------> **80%** **90%** Carbon Neutral*

Today   2020   2025   2030

☐ Fossil Fuel Energy Reduction   ☐ Fossil Fuel Energy Consumption

The 2030 Challenge

Source: ©2015 2030. Inc./Architecture 2030. All Rights Reserved.
*Using no fossil fuel GHG-emitting energy to operate.

FIGURE C.16. The 2030 Challenge for Buildings. Credit: © 2015 2030 Inc./Architecture 2030. All rights reserved.

FIGURE C.17. Flanders Red Cross, Mechelen, Belgium. This new headquarters office was certified at the BREEAM Outstanding level in 2014 by Bopro, a BREEAM Auditor. Credit: © Rode Kruis Vlannderen

FIGURE 5.1. What's in a Green Meal? Credit: Evans Design Studio

Saving Future Operating Costs

Building owners expect that a green building certification will save future energy costs, which tends to be the largest uncontrollable cost of building operations. If green building has to hang its hat on any rack, it has to be on decreasing energy use and offering water savings, now that drought is more or less a permanent condition in the western United States LEED-certified new buildings overall do save energy compared with conventional buildings, and so achieving these savings has become a significant reason for formal certification.

However, annual energy costs for a typical US building are less than about $2.50/sq. ft.; saving 25 percent amounts to only $0.63 (or less) per sq. ft. each year, not a huge amount compared with the cost of certification.

Rich Michal is Executive Director and Chief Facilities Officer at Butler University in Indianapolis, Indiana, which signed the American College and University Presidents' Climate Commitment (ACUPCC) in 2012 and has committed since 2009 to build its new construction projects to achieve LEED Silver. He acknowledges that while campus sustainability has green building as a key component, operating cost savings are critical.

> The nice thing about Butler is we've been committed to sustainability for a long time. I think it's important not to get hung up on rating systems; we try to look at things from a life-cycle standpoint. Part of that is just being pragmatic, because I've got a staff of 100 folks; our 40 campus buildings total about 2.1 million sq. ft. The pragmatic side is that we're not only worried about the first cost of building—we recognize we have to maintain these buildings in the future. Butler University was founded in 1855. We've been on this campus since 1929, and I don't think we're going to move.
>
> We know the cost of maintaining our buildings, and we want to make sure that as we're designing these buildings we're thinking of the life-cycle cost. We want to make sure that we maintain that balance between keeping the first cost under control, but not at the expense of the long-term sustainability of the building. So, we look at sustainable building more globally; it's not just about carbon-emissions; those are important, even critical. But for us, just as important are the long-term life-cycle cost and the functional operation of campus buildings.[7]

Providing Credibility for Sustainability Reporting and Green Claims
LEED's core strength and perhaps its crowning achievement is that it has become a "brand," a shorthand indicator for sustainable achievement. As a reporting system, most people in the building industry accept that it indicates a building (or space within a building) is built or operated according to higher environmental standards. This is an

intangible but real benefit, but only to certain owners. Who benefits the most?

Commercial Real Estate: There is little doubt that a LEED label on a building helps to sell commercial office projects to tenants. Several academic studies show that a LEED building can command higher rents and higher resale prices, and there is also evidence that such buildings lease up faster.[8] In addition, many institutional investors such as public pension funds have explicit real estate investment policies that require a LEED label.

USGBC's President, Roger Platt, puts the business case for this market segment into four words: "Our Investors Require It." He says,

> I was first introduced to the growing investor preference for certified green building projects in Stockholm [in 2010]. I asked a market expert why so many buildings were LEED-certified in a country where eco-construction techniques exist in abundance. He told me that more than 50 percent of commercial buildings in Scandinavia have international investors, and they are requiring LEED. Why? Because they want an international benchmark for their global portfolios.[9]

Large Corporations: Many large corporations use LEED for at least some projects; according to a recent survey sponsored by USGBC, among the 200 largest US companies "Eighty percent agree that LEED is a key way their company communicates sustainability efforts to stakeholders."[10]

Universities and Colleges: Higher education institutions are in constant competition for students, and having a "green" campus is one way to merit attention from those with strong environmental values. Green campuses' annual rankings are a staple of college recruiting. Having many LEED-certified buildings helps to secure such a ranking.[11] The "power users" of LEED tend to be a relatively small group of larger public universities, which require LEED certification of new construction for policy reasons, and elite private universities such as Harvard, which certified its 100th LEED project in 2015.[12] That said, only 523 new higher education projects registered under LEED in 2014 (Figure 8.4) and the total represents *only one project for about every nine* US higher-education institutions.[13]

Responding to Government Incentives and Policies

Since the early 2000s, USGBC and its chapters have effectively lobbied state and local government to provide incentives for LEED certification, especially in new construction. For example, the State of Washington requires all new state-funded construction, including at public community colleges and universities, to be certified LEED Silver. Arizona State University requires LEED Gold certification for its new construction projects. Some cities offer priority processing of building permits to projects that are registered for LEED or Green Globes certification. The State of Nevada offers 35 percent property tax abatements for five years to projects certified at LEED Silver (or Two Green Globes). (For a $500 million casino project, that's real money!) The list is long and constantly changing.[14]

The sticking point is this: For noneconomic initiatives to enjoy widespread support, green building certification systems must be clearly understood, reasonably priced, valued by users, *and* deliver projected operating cost savings. Otherwise, it can be a hard sell to politicians, agency or facility management staff, and the public that a green building should cost more than a minimal amount (extra) to get certified and that certification has value beyond the label.

From the beginning, government has been a major supporter of LEED certification. The public and nonprofit sector is mostly *policy-driven* in making green building decisions. However, in government, it is common practice to separate capital construction costs from future operating costs, so it is often hard to argue for spending more upfront in measures designed to reduce future energy costs unless there is an explicit policy supporting green building.

At the federal level, the Obama Administration has focused its green building policies on reducing future operating costs through an aggressive push to cut energy costs and reduce greenhouse gas emissions.[15]

The Business Case for Green Homes

The market for green home certification is almost exclusively for new homes. Green home certification programs, including LEED for Homes and the National Green Building Standard (NAHB Green, based on the International Code Council's *ICC 700* national standard), as well as strong regional brands such as Earth Advantage in the Pacific Northwest and EarthCraft Homes in the Southeast, aim to help builders sell homes.

A 2009 study analyzing 10,000 third-party certified new homes in Seattle and Portland markets showed that in the Seattle metro area such homes sold at a 9.6 percent price premium when compared to similar noncertified homes, and in the Portland metro area they captured a price premium between 3 and 5 percent. In addition, certified homes sold, on average, 18 days faster than noncertified homes.[16] A 2014 follow-up study for green home developments in several US regions showed similar results.[17]

McGraw-Hill studied US green homebuilding in 2014 and found that 19 percent of single-family homebuilders already build 90 percent or more of their projects to green standards (although their "green standards" definition is much looser than actual certification), with that number expected to double by 2018.

According to this survey, 13 percent of single-family homebuilders and 32 percent of multifamily homebuilders used LEED for Homes, while 30 percent of single-family homebuilders but only 13 percent of multifamily builders used the NAHB Green/ICC 700 standard, almost the exact opposite.[18]

The NAHB green standard is supported by the single-family homebuilders' national association, so it should be more prominent in single-family, whereas LEED for Homes had a much more active multifamily program during the 2010–2014 period, one that cut costs and red tape dramatically for apartment and condominium developers.

Public Relations and Marketing

Clearly, at one time "the first kid on the block" to have a LEED certification had some public relations and marketing benefit. This may still be true for green homes in attracting local media attention and interest from consumers who prefer green homes. One 2014 survey showed that 73 percent of single-family homebuilders and 68 percent of multifamily home builders believe that homebuyers would pay more for green homes (the main reason is to save money on energy)[19] and the easiest way to reach potential buyers is through public relations and marketing.

Employees—The Elephant in the Room

US commercial building energy costs average less than $2.50/sq. ft., while downtown Class A office rents (or amortized construction costs) are about $30 to $40/sq. ft. (or more), depending on the city.[20] However, if employees cost on average, say, $80,000 per year, counting salary and benefits, and if they occupy 200 sq. ft. in an office building on average, then each employee costs $400/sq. ft.[21] In the modern office, the trend is toward less area per employee, sometimes as little as 100 sq. ft., as working modes change more toward collaborative teams that are often put together in clusters, so employee costs per sq. ft. would be even greater. Figure 5.2 shows the relative importance among energy, rent and employee costs, and shows why improvements in productivity are so important to building owners and occupants.[22]

At $400/sq. ft. employee cost, it's easy to see that green design features that offer benefits such as daylighting, views to the outdoors and enhanced indoor air quality are critical to economic performance. Even a one percent gain in productivity would be worth $4/sq. ft., an amount nearly double the annual energy bill for most companies!

A study by University of San Diego and CB Richard Ellis (CBRE) found that tenants in green buildings were more productive and needed fewer sick days.[23] In addition, the study found a reduction in sick days that translated into nearly a $5/sq. ft. benefit, and an increase

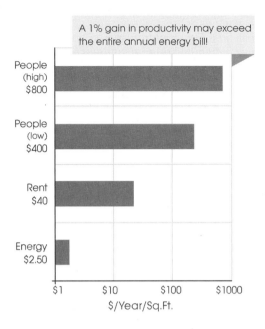

FIGURE 5.2. Why Productivity Improvements Are So Important vs. Energy Savings

in productivity that resulted in a net benefit of about $20/sq. ft.—in other words, these green features provided benefits close to a tenant's total rent and eight times greater than the energy cost! Over six years, a company could expect that a new green building's entire cost (at $150/sq. ft.) could be paid for with just an increase in productivity and a reduction in sick leaves!

A much-ballyhooed recent study purported to show that "green" buildings significantly improved cognitive function of office workers.[24] On closer examination, the study specifically dealt only with controlled ventilation rates, carbon dioxide concentrations and VOC levels in a test environment, not with actual indoor air quality conditions in LEED-certified buildings. Indoor air quality is an important issue for worker productivity, to be sure, but one can still certify a building at LEED Platinum that does not replicate the air quality levels used in the study (except for ventilation rates which are code in many areas and a LEED prerequisite).

Summary

Most companies are skeptical that LEED's benefits will be realized in practice and tend to discount gains in productivity and reductions in sick leave when making investment decisions for buildings. Therefore, the best way to sell green building certification to the private sector is to stick to four key drivers:

1. Provide a menu of options for green construction and operations.
2. Save future operating costs, especially for energy and water.
3. Ensure credibility in sustainability reporting.
4. Take advantage of government incentives.

PART II

GREEN BUILDING HITS THE WALL

CHAPTER 6

Successes: Positive Impacts of LEED and Other Rating Systems

How do I love thee? Let me count the ways.

⋮ Elizabeth Barrett Browning[1] ⋮

LEED, BREEAM and other rating systems share a history replete with success.[2] Formal, technically sound and widely supported green building rating systems, introduced in the 1990s and early 2000s, forced architects, engineers, contractors, building owners and developers to acknowledge the energy and environmental implications of their new building design and construction practices. Later on, as rating systems expanded to cover existing building operations, they sought to provide the same function: to convince owners to deal with issues related to energy use, water use and environmental impacts from ongoing operations.

By focusing not just on energy performance, but rather on much broader design issues, LEED and other systems effectively prompted building teams to take an integrated approach to building design and construction and gave building and facility managers tools to more fully account for operational impacts. They gave everyone a common definition for a "green building." Finally, LEED and BREEAM have issued certifications for more than 40,000 commercial buildings and hundreds of thousands of homes and apartments. Certainly, that deserves to be called a resounding success!

Defining Green

Before the 1990s, a green building had no standard definition and therefore gave no guidance to design teams for making their buildings green. Kim Shinn is a mechanical engineer, LEED Fellow and principal for a leading building engineering firm in the Southeast. In his view, LEED provides three leading benefits:[3]

1. LEED gives consensus-based, industry-led, bottom-up "best-practices" guidance for project teams.
2. LEED created opportunities to expand the discussion about the value and benefit of better building performance—not just energy efficiency, but also better indoor environmental quality and decreased environmental impact. This arguably has led to greater rigor in building codes (i.e., "raising the floor").
3. LEED provides independent, third-party certification and definition of high-performance building [design] and [operational] performance.

I think most people would agree that LEED represented a major step forward for green building in the United States. In Shinn's view, several specific elements tracked by green building rating systems have resulted in major advances in how architects, engineers and contractors approach their work:

1. Higher minimum energy-efficiency standards and greater rigor in energy conservation building codes
2. Better building construction practices—reduced environmental impact, especially in reducing construction and demolition waste going to landfill, as well as increased demand for locally-sourced building materials
3. Building commissioning's rise as a quality-improvement process for the industry
4. Environmentally preferable products, e.g., low-VOC paints and adhesives.

Alan Scott, an LEED Fellow and an AIA Fellow, says,

> I think the big benefit of LEED is that it has raised awareness; terms like LEED and Green Building have become house-

hold words that people can understand, whereas I used to get blank stares 15 years ago. LEED has increased awareness about the issues, even if people haven't fully embraced it and taken substantive action. The other piece I think has been amazing is that, although certified buildings still make up a relatively small chunk of the total population of buildings, LEED has prompted a lot of innovation in materials and technology and pushed manufacturers to stay ahead of the curve.[4]

Other LEED Fellows we interviewed also pointed out that a commitment to certification, coupled with requirements for documentation and third-party review, provides a greater certainty that green building design measures will be implemented during the often-messy construction process.

Integrated Design

Integrated design has had perhaps the greatest impact on the architecture profession, representing a holistic approach that forces architects to bring in more wide-ranging considerations, engaging engineers, contractors and others much earlier during the design process.

LEED and integrated design were not always easy to sell to architects: I recall a meeting in 2001 with a leading architectural firm in Portland, OR, when Alan Scott and I presented LEED version 2.0 to a roomful of staff. We were peppered with tough questions by the firm's principal, who was already successful in designing large office buildings and major corporate campuses. At that time, most architects were unwilling to take guidance on design issues from those outside the profession and wanted good reasons for including green building features in their designs.

In the beginning, LEED was revolutionary because it challenged that viewpoint, pointing out that energy and environmental impacts of buildings were too important in the larger societal framework to leave their design and operation solely to self-interested parties, namely architects, engineers, contractors and owners. During the next few years, that radical viewpoint took hold so strongly that the architectural profession, represented by the American Institute of Architects, essentially capitulated in 2005 and began to work with both USGBC and ASHRAE to improve LEED.[5] The energy and environmental barbarians at the gate[6] were eventually admitted to the architectural world's polite and (very often) self-referential discourse.[7]

By 2008, integrated design had developed into a fairly recognizable system, one chronicled by various authors.[8] While not always observed in practice, integrated design methods are now included for credit in most green building rating systems.

Three recently completed federal buildings showcased how those projects' integrated teams delivered greener buildings.[9] One of these, Federal Center South in Seattle (Figure C.4), completed in 2013, focused particularly on documenting energy use during the first year of operations and even tied a small amount of project team compensation directly to meeting energy performance goals.

Professional Education

By accrediting about 200,000 industry participants as LEED APs and LEED GAs, USGBC and LEED contributed significantly to the industry adopting better design and operations techniques and to new ideas getting a faster uptake. LEED Accredited Professionals must attend continuing education courses and other educational activities to maintain their credentials.

This professional cohort is large: membership in the American Institute of Architects is about 85,000[10] and membership in ASHRAE, the mechanical engineer's society, is about 50,000 (worldwide). Thus, LEED APs and LEED GAs are larger in numbers than all US licensed architects and mechanical engineers.

Through this educational program, early approaches to green design were disseminated rapidly; new ideas and new rating systems gained currency much faster.

Better Energy Performance

Any (and every) green building rating system aims to improve building energy performance beyond that which would have occurred in its absence, i.e., a building built to existing building code requirements. For example, 121 US LEED-certified buildings, built between 2000 and 2006 and studied in 2008, showed an *average* performance *28 percent better* than a "code" (or conventional) building (compared with 25 percent modeled energy savings for the same buildings).[11] The study also revealed that outcomes ranged broadly, from buildings that cut energy use more than 50 percent, to 25 percent that used *more* energy than an average building. This result led many LEED critics to question how a building could be certified with such poor energy performance, a critique that continues to detract from LEED certification's image even today.[12]

It seems fairly obvious that a LEED building that performs as designed will use less energy than one that does not have to meet that standard, yet "the proof of the pudding is in the eating." Energy models may be built on assumptions that do not reflect actual building operations.

Building Commissioning

By requiring that all LEED-certified new buildings undergo a "commissioning" process according to established protocols, LEED supports better energy performance as a goal. Studies at Lawrence Berkeley National Laboratory as early as 2005 documented that properly commissioned buildings were likely to reduce energy use by 10 to 15 percent compared with similar buildings that were not commissioned.[13] So far, so good, but buildings are like pianos: they require regular tuning through "retro-commissioning"; therefore, it's important that all new buildings commit to using this practice to maintain their energy performance levels.

One expert attributes LEED certification's value, particularly in new construction, primarily to the commissioning process, the documentation required and the requisite third-party oversight:

The highest value proposition of LEED certification isn't the plaque; it's the third-party independent verification. There is a cost associated with in-depth, multi-disciplinary technical evaluations of buildings, which are, for the most part, custom-designed, custom-built and highly complex. Without third-party verification, there is no way to know that what was intended was accomplished. Those who plan to design to LEED standards but save money by not getting certified miss the essential benefit of certification, which is the affirmation of [meeting] the standards. Without affirmation, you won't know what you've ended up with, but you can pretty well expect it not to perform as intended, outstripping any savings realized by skipping certification.[14]

Better Products

While building products constantly evolve according to technological opportunities and changing customer and designer preferences, it seems clear that LEED has had an important role in accelerating building products' rapid evolution, through creating standards that forced manufacturers to compete based on greener product features. A few examples come to mind:

1. Low-VOC products.[15]
2. Bio-based materials.
3. Recycled-content materials.
4. Low-water-using toilets and urinals.

LEED's inclusion of more restrictive California or Green Seal standards for VOCs gave manufacturers an impetus to create innovative products that eliminated "new building smell" from LEED buildings. It turned out, for the most part, to be easy and cheap to create these new products because they were already required by air pollution control agencies in California, the largest single US market for building products. What happens in California (usually) doesn't stay in California, and products such as zero-VOC paints are now widely available.

However, in some cases, LEED requirements were written without knowing much about how building products manufacturing works in the United States. Kimberly Hosken is a LEED Fellow with more than 30 years' experience in construction. She says,

> In early design charrettes in California, we would hear team members say, "We've got to get the steel from less than 500 miles away so we can get this LEED point." and I'd say, "Steel is produced at specific locations in the United States, the majority of these facilities are on the East Coast. There are a limited number of steel production facilities in this part of the country; it's a commodity. You can't control where steel comes from through specifications, and the construction procurement process is not set up to be able to do that without adding significant cost and potential delays. It is a complex process which requires a dialog for what you can or should try to control through specifications."

In addition to steel, a majority of building materials are historically manufactured in locations that were based on economics. The textile and furniture industries started in the eastern part of the United States, with cotton grown in the South and lumber from forests along the East Coast. Products were manufactured on the eastern seaboard with the earliest mills in New Jersey and Rhode Island. Are there exceptions? Sure, but the majority of textiles [still] come from the South. So, this is the infrastructure of manufacturing in the United States, and LEED has never acknowledged that. They just say, "Get your products from local sources." They don't acknowledge that there's this 200-year-old history of why things come from where they come from. However, the great benefit of LEED is that we had these conversations and we started to learn where products came from, what they were made of and how they were made.[16]

Occupant Comfort and Satisfaction

Many studies have found higher occupant satisfaction levels with LEED-certified buildings. A 2011 federal GSA study of 22 buildings either LEED-certified or built with green design principles showed 76 percent higher occupant satisfaction.[17] Other studies have found significant correlations between occupant satisfaction and design parameters such as daylighting, views to the outdoors and improved indoor air quality, all encouraged in both LEED buildings and in standard sustainable design practice.

Government Policy

While initially some government agencies took the lead in establishing LEED as their "go to" rating system, by 2006 USGBC and its chapters began to take a more proactive role and became more active in securing government support in both policy and law. Since then, they have been effective lobbyists to secure incentives, regulations and policies to support LEED, often excluding other rating systems such as Green Globes.[18] While I support some government action to encourage green building, there has been limited accountability to ensure that finished projects demonstrate that they meet objectives for significantly lower energy use and reduced overall carbon emissions.

Soon we may see green building codes gaining widespread adoption with similar poor accountability, as USGBC, ASHRAE, AIA and International Green Construction Code (IgCC) join forces to promote both regulation (adopting the IgCC) and "voluntary leadership."[19]

Commercial Successes

As we've seen, LEED works well for several market segments, e.g., downtown office developers in large cities, where owners can monetize the LEED label and get significantly greater returns by investing in LEED certification, and also gain a competitive advantage over similar non-LEED buildings nearby. For this reason, green building's benefits have become widely accepted by this ownership group since about 2006.[20] Table 4.1 shows commercial (nonresidential) buildings certified by LEED in 2014. Certifying more than 3,350 projects and 417 million sq. ft. in the United States (4,405 projects and 657 million sq. ft., if one counts international projects as well) is a significant success and something that USGBC should rightly celebrate.

The 2015 National Green Building Adoption Index (NGBAI) documents that about 13 percent of office buildings in 30 major US metropolitan areas (representing 39 percent of the total area of commercial offices) have achieved LEED or Energy Star certification, with 5.3 percent LEED-certified (representing 20.3 percent of all commercial office area).[21]

David Pogue is CBRE's Global Director of Corporate Responsibility, the sponsor of NGBAI; he is a long-time sustainable building and operations proponent. Regarding LEED's successes and future challenges, he says:

> I think that LEED has put a bright spotlight on those elements of buildings that impact environment and people, and I think they have changed the marketplace forever. They have moved the entire market to higher standards. And importantly they have demonstrated that you can build better at almost no additional cost. USGBC should be quite proud of what they have done in that regard; high-performance and sustainable practices are now an accepted and, in many markets, a required aspect of being a Class A office building.
>
> But there is much more work to be done. I believe that one of their original goals was to use those same standards to improve the broader market, but I don't think there's been as much "lifting of all boats" as hoped. This is due to many factors: part market demand, part knowledge gap and part economics. There may still be a perception among certain uninformed owners that "It's too expensive. It's too difficult, so I won't even try." As we saw in the NGBAI study recently published, there's very low participation among smaller buildings, and that's the biggest part of the building stock.

I think that the upper end has dramatically improved, and I suspect that, over time with changes in building codes and the like, we will improve the lower end. However, that will take a great deal of time and that is a problem.[22]

In Chapter 5, we discussed LEED's proven benefit in getting higher rents, faster leasing and higher resale value for large commercial office buildings in large cities' downtown areas. That's great for those buildings, but it leaves out most commercial properties, as well as all other properties that are owner-occupied, including government, university, K12, healthcare, retail and other building types. According to Pogue, larger corporate tenants that tend to rent space in giant buildings are more likely to have sustainability baked into their mission statements.[23]

There are regulatory requirements, there are corporate initiatives, and then there is the ownership side. All of that taken together has propelled the LEED certification-mania over the last five to seven years. It may have started as an opportunity to lead [the market]; now I think it's a requirement not to trail.[24]

The NGBAI reveals that, for properties with less than 100,000 sq. ft., only 4.5 percent have an Energy Star or LEED certification, collectively representing only seven percent of total commercial building space. In this comprehensive study of 35,000 office buildings, for the smaller office buildings in the 10 largest markets, LEED certifications totaled 1.5 percent of the buildings and 1.8 percent of the building area (in 20 smaller markets the numbers were 1.5 percent and 2.3 percent, respectively), findings that are entirely consistent with the analysis in this book.[25]

Pogue agrees, saying:

As we saw in the GBAI study, there's a significant drop-off between the buildings above 250,000 square-feet and the buildings below 100,000 square-feet, for instance. There is very low participation in the smaller buildings, and that's the biggest part of the building stock.

The NGBAI also presents some conclusions that are not good news for LEED:

> The overall results show that the uptake of green building practices in the 30 largest US cities continues to be significant, but the growth shows abatement. [This result] reflects the fact that only a certain fraction of the building stock can obtain a sustainability or energy-efficiency certification. And perhaps it also indicates that *the fraction that can seek certification has now done so.* The most sophisticated owners with the most high-profile buildings in Tier 1 markets have [already] pursued and achieved certification.[26] (Emphasis added)

LEED works for the Fortune 500 because it's become a shorthand label for green building that's easily communicated to employee-associates and other stakeholders. For most large companies, government incentives do not figure significantly in their decision to build to LEED standards and to certify.

The LEED "Ecosystem"

Because it fostered a community among professionals engaged in green design, construction and operations, USGBC and LEED created an ecosystem, shown in Figure 6.1, which helped LEED's use grow rapidly during the 2000s. The ecosystem has seven elements, interwoven into a green building certification tapestry:

1. USGBC members, numbering some 12,000 companies in 2014 (down 40 percent from about 20,000 members at the pre-recession peak). As a membership organization representing companies and not individuals, USGBC has had more marketplace influence than AIA and ASHRAE, which have only individuals for members.

2. Twenty-two LEED rating systems, divided into four broad categories: new construction (and major renovations), interior remodels and fit-outs, building operations, and residential construction.
Each category contains one or more specialty systems for common building types (schools, retail, healthcare) and there is also
a system for planning larger developments. These allow almost
any conceivable building to be rated and certified, from detached
single-family homes to large integrated urban commercial developments.

3. Approximately 200,000 LEED APs and LEED GAs provide consulting expertise for executing LEED projects and create a strong
lobby at the project level to use LEED ratings.

4. LEED has provided public agencies and elected officials interested in promoting green building and operations something
tangible to advocate. As a result, LEED policies, regulations and
incentives have been adopted at federal, state and local government levels across the United States. The same phenomenon has
occurred in other countries, sometimes with LEED, sometimes
with local rating systems.[27]

5. Building owners and developers became key advocates for LEED
in a few market segments, including commercial offices and corporate real estate. Developers in large urban areas found that they
could successfully "monetize" the LEED-certified label (even the
LEED pre-certification label in the LEED CS system).[28]

6. The LEED bureaucracy (including GBCI and USGBC staff) develops rating systems, oversees the LEED technical committees,
publishes the LEED manuals to explain the systems, creates hundreds of addenda to the LEED systems, operates the review system for LEED documentation, develops new rating systems and
adaptations, creates (tens of) thousands of credit interpretation
rulings, maintains a library of hundreds of alternative and pilot
credits, maintains the LEED database, and functions in myriad
other ways as the keeper of the flame.

7. Finally, there are thousands of LEED-registered and certified
projects throughout the United States and internationally that,
taken together, offer a rationale for the LEED system, act to doc-

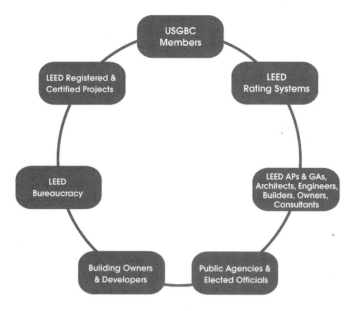

FIGURE 6.1. The LEED Ecosystem[29]

ument achievements and create incentives for an owner's next project to be a LEED project.

Summary

I don't think anyone could question that USGBC and LEED have had a very positive impact on developing and implementing sustainable building design, both in the United States and worldwide. USGBC has aggressively marketed LEED around the world and has helped to jump-start many national green building councils in nearly 100 countries. All this is for the good; however, as we've demonstrated, green building certification in its primary market, the United States, has hit the wall in the past few years and isn't growing on an annual basis.

If we in the green building movement are going to accomplish our stated goal of transforming the built environment and, especially, tackling urgent issues of climate change from building energy use, we need to examine rationally why LEED has stopped growing in the United States and what we can do about it. The rest of the book deals with these two subjects.

CHAPTER 7

Failures:
LEED's Limited Appeal

*Victory has 100 fathers and
defeat is an orphan.*

President John F. Kennedy[1]

In Chapter 6, we recounted many ways in which LEED can be considered a great success. However, achievement in business is best measured by how many people in the total market use your product or service. In that regard, LEED has failed miserably, as we indicated in Chapter 4 and will document extensively in Chapter 8. Although LEED's ecosystem is well established, it too is fraying around the edges, as more people find alternative ways to build green without using LEED.

Moreover, *LEED has spectacularly failed in dealing with the existing buildings market,* where most carbon reduction opportunities are found. In 2014, for example, LEED certified 550 projects under the LEED-EBOM system, exactly 0.01 percent of 5.5 million US nonresidential buildings. *Zero-point-01 percent*—that's not a typo; in 2014, LEED-EBOM certified only one in every 10,000 US buildings (or 0.2 percent of the total nonresidential building area, if you like), representing nearly a complete failure to cut carbon emissions with this rating system. Office and retail buildings were the only users of any consequence, representing 91 percent of all LEED-EBOM certifications during 2014.[2]

For some, LEED may appear to be a quasi-religion, with its own priesthood, rites and rituals, acolytes and catechism; nevertheless, the "un-churched" audience is huge, *representing more than 99 percent of buildings and homes and more than 96 percent of all building area.* (See Figures I.1 and 4.2) Let's say it once again: *More than 99 percent of US buildings are not LEED-certified.* Most organizations would consider that as an amazing market opportunity; for me, after 15 years of nonstop promotion and hard work by USGBC and tens of thousands of professionals, it indicts the LEED system and calls into question USGBC's entire approach to the market.

LEED Fatigue

Even in successful markets, "LEED fatigue" has set in: Clients don't want to pay extra for a certification that by itself may not add value to a project. Michael Deane is a LEED Fellow and head of sustainability for Turner Construction, the largest commercial builder in the United States. He says,

> There seems to be a decline in people who are going for formal certification, but they still claim that they're building sustainably and/or that they're following the LEED playbook.[3]

Chris Forney of Brightworks in Portland, OR, is an experienced LEED consultant. Of this phenomenon, he says:

> We have clients all the time that ask about going through and using LEED as a guide, [but one] that [they will] not certify to—it's commonly referred to as "LEED Lite," and in different client scenarios it's been a reasonable way to go. One example would be with earlier versions of LEED, like LEED version 3; one client in particular wasn't seeing a lot of value in going after the materials points, because he was just looking at regional materials and rapidly renewable materials. This project is a simple building type; they would typically get both the points—20 percent regional materials and 20 percent recyclable content—but the effort it took to

compile that data for each project…was just more work on the part of their contractors, and the tracking wasn't creating any additional benefit for the project.[4]

In some ways, LEED has suffered the worst fate of all rating systems; it is used by many designers and building operators as a guide to sustainable design and to building operations, but is not regarded as important enough to pay for a certification. Many building teams say they are "designing to LEED" but don't bother to register or certify their projects; this seems like living together without getting married, and so this approach lacks one thing essential for a well-functioning building (or relationship): *commitment*. Project teams register projects under LEED to please clients and satisfy onlookers, without ever planning to certify the project. GBCI collects the rather small registration fee, but has much good really been accomplished?

Why is this? LEED is widely seen as too expensive by most building owners, as unneeded guidance by many architects, and even as outdated by others. Indeed, in mid-2014 USGBC extended the date for mandatory registration of new projects in LEEDv4 until October 2016, from its original mandatory date of mid-2015, a move that was widely derided in the LEED blogosphere. That decision will allow projects that register before October 2016 under the LEED 2009 system to be certified, as long as they finish documentation by mid-2021 (Figure 7.1). In other words, projects can be certified as "green" as late as 2022 by incorporating sustainability approaches that were the "state of the art" in 2009, when LEED 2009 first came on the market.

When I first wrote in 2009 about certifying LEED for existing buildings,[5] LEED consultants cost typically about $50,000 to help an owner certify a building, added to a company's in-house team's efforts to assemble data. Even at that cost, many companies saw certification as an expensive indulgence, except those engaged in commercial real estate, which, as we saw in Chapters 5 and 6, could easily monetize the investment. For owner-operated buildings, which don't compete for tenants, LEED-EBOM has begun to exhibit the faults of other LEED

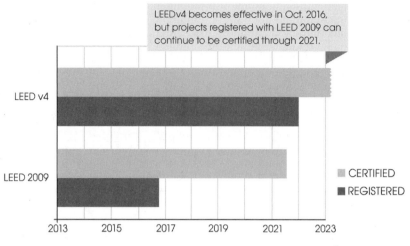

FIGURE 7.1. Effective Dates for LEED 2009 and LEEDv4[6]

rating systems: it is expensive, bureaucratic and virtually useless as a management tool to achieve green operations on an ongoing basis.

Designed to LEED

Many projects advertise themselves as "Designed to Achieve LEED Silver" or "Registered for LEED Silver," designations that don't exist in LEED. As a result, the marketplace has come to accept that a project can be designed to be "LEED certifiable" without ever going through the rigor and cost of certification.

Consequently, many projects and clients are questioning why they should proceed with LEED certification, even when they have major commitments to green design and sustainable building operations. Curtis Slife says:

> We do a considerable amount of work with municipalities and large corporations, and almost all of them are adopting a "LEED Certifiable" approach, instead of going through the expensive and time-consuming LEED Certification process.[7]

More than just not wanting to pay to have someone certify what it's already doing, if a company just wants the PR benefit of green building, they can advertise, as many do, as "LEED Registered" without ever planning to pursue certification. Think about it: for the $1,200 price of registration, a new construction project, a building renovation, or an interior remodel or fit-out can secure many benefits of certification, without having to incur the cost or wait for final approval—who could blame them?

In a way, you could say this is a success, that LEED has become so widespread that it's become generic for "green building," much as Kleenex became generic for "tissue" and Band-Aid for "bandage." In the private sector, however, marketers and companies fight for years to prevent their brand names from becoming generic.

Let's look at LEED's disappointments in a more systematic way. There are three main symptoms of failure:

1. LEED has not "transformed" the US building market or the built environment. As we demonstrated in Chapter 4 using LEED's project data, LEED has certified less than four percent of US commercial building area after 15 years of hard work.

2. LEED is "an inch wide and a mile deep." It simply does not work for a large number of important market segments outside of commercial offices and high-end corporate and government buildings, because total cost is much greater than perceived benefit.

3. LEED's customer experience is questionable. Projects cost too much, requirements are hard to understand for many project teams, the value of the LEED label is lost on most companies, and the system doesn't deliver certification results in a reasonable period of time.[8]

Let's examine each of these failures and listen to some experts who have been using the system for a long time.

LEED Has Not Transformed the Built Environment

Even for large corporations with strong sustainability goals and very sophisticated operational practices, often LEED just doesn't fit with their way of doing business. LEED Fellow and former USGBC staff member Kimberly Hosken says:

In the past six years [while at a Fortune 100 company] there was an appetite for LEED on a corporate level. However, I was working with mega-customers with hugely complicated sites and global portfolios, and LEED didn't fit well. While [I was] at USGBC my focus was volume certification. I worked with staff and customers to create and implement volume certification, and we figured out how to make it work for all kinds of applications.

When it came to global corporate existing-building portfolios, volume certification was complicated. It was difficult to connect and "sell it" as aligning with corporate sustainability goals. With global portfolios we would benchmark them for LEED and then the customer would say, "Well, we can't do it. We can't meet prerequisites; we don't have meters the way it's required by LEED—we might have a lot of meters, but they're not set up to meet the requirements of LEED (or ENERGY STAR even)." This is why I'm excited about the GRESB (Global Real Estate Sustainability Benchmark) program.[9] Although GRESB doesn't exist for corporate [facilities' use] yet, it is specifically designed to be applied for portfolios.[10]

By any standard, LEED's vision—*businesses and communities will regenerate and sustain the health and vitality of all life within a generation*[11]—cannot be met using the current approach. Yet, one finds the same assertions about LEED's "successes" repeated endlessly to justify the existing approach.

Early on, USGBC staff would show a chart showing 25 percent of buildings as the goal for certification (Figure 4.10), presumably gaining enough critical mass of that the rest would follow suit.[12] That goal has apparently been abandoned, and what we read instead are just frequent USGBC reports and press releases about how many square feet are registered and/or certified each month and year.

In the real world, saying it doesn't make it so. Our detailed analysis in Chapter 8 shows that LEED will not reach its goals fast enough to make a real difference by 2030 and therefore green building certification has become almost irrelevant to global goals for reducing carbon emissions from the built environment.

LEED's Real Appeal Is Quite Limited

As we showed in Table 4.1, in 2014 LEED registered about 277 million sq. ft., about 15 percent, of new commercial construction (excluding commercial interiors projects) but only about 0.2 percent of existing building area. That's a real achievement, but still relatively limited. At year-end 2015, LEED had certified just 0.8 percent of *all* US nonresidential buildings and less than four percent of *all* commercial building area. Residential certifications represented even far lower percentages than commercial totals.

In the existing building market, LEED's appeal is *much more limited* (Figure 8.8.). For the period 2010 through 2015, LEED certified fewer than 3,100 existing commercial buildings (out of 5.6 million) in the United States, about *0.06 percent* of the total number.

Let's take a closer look at the US building census, shown in Table 7.1.[13] In 2012, there were 5.6 million nonresidential buildings, occupying about 87.4 billion square feet. Buildings over 100,000 sq. ft. represent one-third (34 percent) of all building area; most buildings are small, however, with a median size of 5,100 sq. ft. Half of all US commercial buildings are 50,000 sq. ft. or less, and that is where LEED is least cost-effective, primarily owing to fixed costs of certification, which lead to high costs per square foot.

The largest single building type is offices, with 18 percent of total area and 1.01 million total buildings; their median area is only 5,000 sq. ft. Education (K12 and college/university), with 389,000 buildings, represents 14 percent of commercial space. Including food sales, retail buildings make up 14.5 percent of all building area, with 779,000 buildings.

In Chapter 8, you will see that LEED has almost no significant following in the education, retail, healthcare, industrial and other markets, which constitute well over 75 percent of the total number of US nonresidential buildings. *Conclusion: No matter how you look at it, LEED is a niche product without much appeal outside its current markets.*

TABLE 7.1. US Building Stock, 2012[18]

	Number of buildings (thousand)	Total floorspace (million square feet)	Median square feet per building (thousand)
All buildings	5,557	87,359	5.1
Building floorspace (square feet)			
1,001 to 5,000	2,772	8,022	2.8
5,001 to 10,000	1,229	8,887	7.0
10,001 to 25,000	885	14,218	15.0
25,001 to 50,000	336	12,015	35.0
50,001 to 100,000	200	13,988	67.0
100,001 to 200,000	90	12,410	133.7
200,001 to 500,000	38	10,767	264.0
Over 500,000	8	7,051	741.0
Principal building activity			
Education	389	12,285	10.0
Food sales	177	1,303	3.2
Food service	380	1,832	3.6
Health care	157	4,151	5.0
Inpatient	10	2,373	101.0
Outpatient	147	1,778	5.0
Lodging	158	5,904	14.1
Mercantile	602	11,425	6.6
Retail (other than mall)	438	5,522	5.0
Enclosed and strip malls	164	5,903	13.6
Office	1,012	16,002	5.0
Public assembly	352	5,539	5.4
Public order and safety	84	1,441	6.1
Religious worship	412	4,562	5.9
Service	619	4,612	4.0
Warehouse and storage	796	13,052	5.5
Other	125	1,995	6.0
Vacant	296	3,256	4.3

LEED's Customer Experience Is Questionable

Larry Clark, with Sustainable Performance Solutions in Fort Lauderdale, FL, has extensive experience as a consultant using both LEED and Green Globes. Of LEED's drawbacks, he says:

> The easily identifiable costs are a major consideration, but the additional costs of meeting hurdles discovered during the course of the project are even more off-putting... and owners talk to each other! With LEED in particular, the delay in resolving comments from the review team can be lengthy, frustrating and burdensome.[14]

Hernando Miranda is a professional engineer, architect and owner at Soltierra, Inc., in Dana Point, CA, heavily involved in the development of several LEED versions and in certifying projects during the past 15 years. He has deep concerns with how LEED works, saying:

> There is excessive third-party review effort [by LEED's review teams]. The professional standard of care that LEED reviewers demand allows only those with insider knowledge to get through a review with minimal effort. In addition, there is a lack of consistent documentation rules; rules need to be locked in place during the time a version of a standard is in use, the same way building codes are. LEED has not been streamlined to simplify the documentation requirements for green measures that have become standard. In addition, LEED does not give sufficient credit to reducing materials use and LEED does not give sufficient credit to zero net energy buildings.[15]

Douglas Carney is an architect, senior vice-president at Children's Hospital in Philadelphia and a faculty member at Drexel University. His concerns about the LEED certification process are:

The process is cumbersome, secret and slow, and you don't know who is evaluating your project. If I were an architect, I would be worried about peers/competitors evaluating my submissions and having less than desirable motivations.[16]

For many years, Stuart Kaplow, an attorney in Towson, MD, has followed and commented on LEED, as well as represented companies engaged with LEED and the construction industry. He cites three main drawbacks:[17]

- The time, inconvenience and expense can be problematic.
- Innovation and creativity are stifled.
- The system is too forced, resulting in homogenous buildings.

The implicit presumption in LEED is that we can, once and for all time, define a green building and that if you want our label, you'll do it our way. I believe American architects, engineers, builders and building owners are too innovative (and buildings are too diverse) to take such guidance from any single system over the long term.

The bottom line is this: each LEED version encapsulates ossified notions about green building design that become more out-of-date and incapable of accommodating changes in professional practice and building products as years go by. The solution, which we detail in later chapters, is not to tinker with individual requirements, but to commit to a wholesale redesign of both the system and the delivery model.

Summary

Despite many well-publicized successes, LEED has failed to transform the built environment in substantial ways, owing to failures in both substance and process. In the next chapter, we present more detailed analyses that document how LEED has failed to secure a position in many large US market segments. At some point, any business (and USGBC is run like a business) that fails to get beyond the "innovator" market (defined as about three percent of the total addressable market) would reconsider its product line and how it goes to market. Shouldn't USGBC do this?

CHAPTER 8

LEED Fails to Transform
the Marketplace

Houston, we have a problem!

Apollo 13[1]

In Chapter 4, we showed LEED's growth flat-lining in the United States since 2010, with LEED losing support in both new construction and existing building certifications. In this chapter, we'll take a deeper dive into LEED project data.

Only by acknowledging what's happening in the marketplace is it possible to start thinking about *why* this is the case and *what can be done* to advance green building. We'll focus on LEED because it has most of the certification market and by far the largest following in the United States.[2]

Overall LEED Project Trends

Figure 8.1 shows LEED's growth and leveling off in the US non-residential market in three 5-year time periods, reflecting its 15-year history to date. From 2000 through 2004, LEED got started, it attracted the innovator market segment and showed the classic "diffusion of innovation" pattern, as it began to trace out an S-shaped or "logistics" market adoption curve, one described in my 2007 green building marketing book.[3]

From 2005 through 2009, growth became much more rapid than would be expected by diffusion alone, exhibiting the "network effect," in which each new project became cheaper and easier to execute because rapid learning took place among architects, engineers and builders; in addition, LEED's brand gained greater acceptance.

During that period, LEED's brand became more valued in commercial and corporate office market segments, and this helped spur growth. However, after the Global Financial Crisis, LEED's use in new construction diminished rapidly, replaced for a few years by a growing market for certifying existing buildings. By 2012 that market also began a slow reduction in annual numbers for LEED-registered and LEED-certified buildings.

The number of LEED-certified projects grew during the next five-year period (2010–2014) while new project registrations didn't increase. Certifications grew because of several factors: the international market's growth (from less than 10 percent to 35 percent of all new LEED project registrations), the previous increase in project registrations that came to fruition during this period, the new construction market's return to growth (especially with larger US office buildings, where LEED has always been strong), the large number of LEED accredited professionals promoting the system, and the strongly held belief in the LEED professional community that this was the best and most realistic choice for certifying green building. Yet in 2015, LEED US nonresidential project registration and certifi-

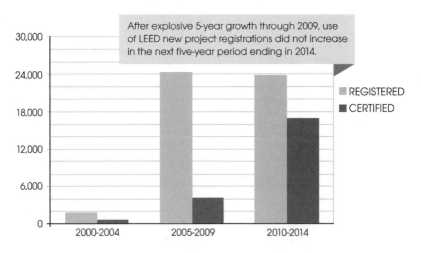

FIGURE 8.1. LEED US Nonresidential Projects, 2000–2014

cation numbers were lower than in 2012 and 2012 to 2014, respectively. (See Appendix, "2015 LEED Projects Update.")

The US Commercial Buildings Market

Table 8.1 shows US buildings by type of use, which helps us to understand where LEED is failing to address the market. More than 67 percent (two-thirds) of US commercial building floor space in 2012 comprised buildings below 100,000 sq. ft., and more than 50 percent comprised those below 50,000 sq. ft.[4] Large spaces, mostly commercial and corporate offices, superstores, hospitals and the like, above 100,000 sq. ft., represented only one-third of total space.

Of US buildings in 2012, 62 percent were built prior to 1990, before the advent of modern energy codes and standards. These buildings need special attention to reduce their carbon footprint.

What's the point? Fixed costs to meet prerequisites for LEED certification, energy modeling (required only for new buildings), building commissioning (required for both new buildings and existing buildings), specific design measures, operational changes, documentation and certification costs, are *prohibitive* for these smaller buildings. To bring the vast majority "into the fold" for green building, *we must reduce certification costs dramatically, by factors of 10 to 100!*

However, it's easy to see why a continued focus on larger buildings is important. Figure 8.2 shows how commercial buildings over 100,000 sq. ft. constitute an outsized share of total floor area, making up 35 percent of total floorspace.

Complicating the picture is that newer buildings tend to be a bit larger than older structures, with better energy use profiles. According to a US government report (CBECS), "Although about 12 percent of commercial buildings (comprising 14 percent of commercial floorspace) were built since 2003, the commercial building stock is still fairly old, with about half of all buildings constructed before 1980."[5]

Clearly, green building rating systems need to be restructured so that older buildings can go through at least an energy-efficiency-based certification process, in order to tackle the larger carbon problem with greater effect.

TABLE 8.1. US Nonresidential Buildings by Type of Use, Number and Floor Area[6]

	Number of buildings (thousand)	Total floorspace (million square feet)	Mean square feet per building (thousand)	Percent of Total Building Area	Percent of Total Buildings
All buildings	5,557	87,359	15.7	100.0%	
Principal building activity					
Office	1,012	16,002	15.8	18.3%	18.2%
Warehouse and storage	796	13,052	16.4	14.9%	14.3%
Education	389	12,285	31.6	14.1%	7.0%
Mercantile	602	11,425	19.0	13.1%	10.8%
Retail (other than mall)	438	5,522	12.6		
Enclosed and strip malls	164	5,903	36.0		
Lodging	158	5,904	37.4	6.8%	2.8%
Public assembly	352	5,539	15.7	6.3%	6.3%
Service	619	4,612	7.5	5.3%	11.1%
Religious worship	412	4,562	11.1	5.2%	7.4%
Health care	157	4,151	26.5	4.8%	2.8%
Inpatient	10	2,373	247.7		
Outpatient	147	1,778	12.1		
Food service	380	1,832	4.8	2.1%	6.8%
Public order and safety	84	1,441	17.2	1.6%	1.5%
Food sales	177	1,303	7.4	1.5%	3.2%
Other	125	1,995	16.0	2.3%	2.2%
Vacant	296	3,256	11.0	3.7%	5.3%
				100%	100%

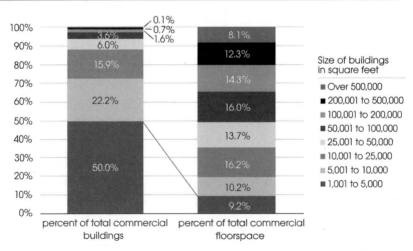

FIGURE 8.2. US Commercial Building Numbers and Floor Area, 2012[7]

Large Commercial and Corporate Offices

As we discussed in Chapter 6, there is no question that LEED has become an accepted brand in commercial real estate, especially in large office buildings in major cities. Most research studies on LEED's economic and financial benefits have focused on this market segment. While important, large commercial offices are only part of the overall building market. Similarly important for LEED is the corporate office market, primarily headquarters and research buildings for large companies.[8]

Also, as we demonstrated in Chapters 5 and 6, the business case for commercial real estate is solid. LEED buildings appear to offer significant benefits to commercial office owners and developers. In this market segment, owners have figured out that a LEED label can help them in the marketplace and, as smart investors, they understand that direct financial benefits outweigh any extra costs.

Smaller Office Buildings

But with smaller office buildings, the situation is markedly different. There are more than a million US office buildings, mostly small, totaling 16 billion sq. ft. of area. Cumulatively, they represent 18 percent of total building area and at least 18 percent of energy use in commercial buildings. LEED has certified roughly 10,000 (mostly large) office buildings, about one billion sq. ft., representing 6.7 percent of total area, but only 1 percent of total number. Considering that LEED certifies fewer than 600 existing buildings per year, it's doubtful LEED will ever certify more than a small percentage of even the most efficient US office buildings.

Public Sector

Early in the 2000s, LEED's uptake in nonresidential buildings relied heavily on the government and nonprofit sectors, exceeding 50 percent of total projects. Beginning in about 2006, the private sector began to dominate LEED project registrations, and in 2014, government projects accounted for only about 15 percent of total LEED project registrations and certifications. However, if we broaden the public sector definition from government alone to include public schools and public higher-education institutions, then the public sector represented 25 percent of LEED project registrations and certifications in 2014, indicating continuing strong support for LEED projects, often for policy reasons.

Education

The education market is huge. LEED has provided certification for both the K12 (elementary and secondary) and higher education (post-secondary) markets from its inception. *That's why a detailed look at LEED registrations and certifications in these markets gives such a surprising result: less than one percent penetration in either market.*

K12 Schools. There are 132,000 US K12 schools.[9] From 2010 through 2015, LEED certified less than one percent.[10] There are good reasons for this. School construction is typically financed with bonds in a process that may begin several years before a school is built; even modest inflation means that school construction is over-budget before it begins, so adding the cost of LEED can be difficult. In addition, the political process of school construction means that it is hard to justify pursuing more costly sustainability objectives unless the school superintendent, the school board and often the capital project manager are all convinced that it is both important and viable within the budget.

Green Schools—A Nonstarter. Despite all USGBC's hoopla for the past five years about the need for "green schools," only a total of 2,738 such projects (2.1 percent of schools) were registered through year-end 2015 and only 1,237 (0.9 percent) were certified (Figure 8.3). At the recent average LEED certification rate of about 200 schools per year (e.g., 193 were certified in 2015), one can see that this task not only will never be finished, but will fall farther behind, given the large volume of school construction going on right now.[11]

USGBC created the Center for Green Schools in 2010,[12] but since that time, use of LEED for Schools has declined precipitously. USGBC's stated goal of "green schools within a generation" will *never* be met at this rate. Some K12 school districts may say that they are us-

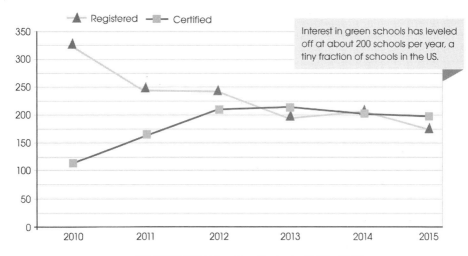

FIGURE 8.3. US LEED K12 Education Projects, 2010–2015

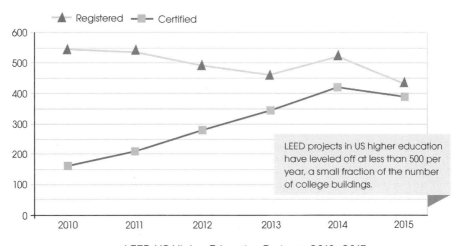

FIGURE 8.4. LEED US Higher Education Projects, 2010–2015

ing LEED criteria in their designs,[13] but without some third-party review, odds are that anything that adds cost will get "value engineered" from the final project.

Higher Education. In 2009, there were 4,500 US higher education institutions, enrolling more than 20 million students.[14] Assuming about 75 students per building, that would total more than 260,000 buildings.[15] From 2010 through 2015, LEED certified fewer than

1,800 higher education buildings (Figure 8.4), representing less than 0.7 percent of total higher education buildings.[16] Registered project numbers have ranged between 450 and 550 per year since 2010, with only 422 projects registered (and 393 certified) in 2015. In other words, there has been basically *no growth* in LEED's use in higher education over the past half-decade.

Consider that 685 college and university presidents have signed the American College and University Presidents Climate Commitment and 539 colleges and universities have submitted Climate Action Plans,[17] and you might conclude that there is only a weak connection in the minds of most universities between using LEED and their choices for carbon reduction initiatives. *The bottom line: LEED is only marginally relevant to higher education's carbon emission reduction efforts.*

Retail

Retail is a huge industry everywhere. In 2014, in the United States, commercial, non-office, non-hospitality construction amounted to $57 billion, second only to education. There are about 1.16 million retail establishments, including freestanding retail stores, shops in malls, food sales (grocery) and food service (restaurants).

In 2014, LEED retail registrations and certifications amounted to a tiny fraction of stores, as shown in Table 8.2. With only 1,500 LEED-certified *new construction* retail projects during the past five years, LEED represents less than 0.2 percent of stores. Assuming each retail project cost $2 million (20,000 sq. ft. × $100/sq. ft.), an average of 300 projects per year would represent $0.6 billion in construction value, or about one percent of the retail new construction market (calculated from Table 4.2).

A few chains such as Kohl's, Starbucks (only 500 out of 12,000 company-owned stores), PNC Bank (about 100 branches) and Stop & Shop grocery stores represent "power users" of LEED, but overall penetration of the retail space is insignificant. USGBC's "LEED in Motion: Retail" report lays it out.[18] Here are some success stories:[19]

- Starbucks currently has 500 certified stores in 18 countries.[20]
- Kohl's is committed to achieving LEED certification for all new

stores. The company has certified 434 buildings, or 38 percent of its stores. Of their total certifications, 286 are for existing buildings.

- Verizon Wireless has 200 certifications, all LEED Silver or higher, about ten percent of their 2,300 locations.[21]

TABLE 8.2. Retail US LEED-NC Projects, 2010–2014 [22]

Year	Registered		Certified	
	No. of Projects	Avg. Gross Area (sq. ft.)	No. of Projects	Avg. Gross Area (sq. ft.)
2010	293	18,288	127	40,508
2011	436	23,470	184	33,645
2012	487	26,256	324	16,087
2013	432	21,894	416	14,479
2014	583	12,110	469	10,498
Total	2,231	20,404	1,520	23,043

The growth of US retail projects is shown in Figure 8.5. Through year-end 2015, fewer than 3,500 US retail projects were certified, and there is continued slow growth in this segment. *The basic conclusion: given the small percentage of this sector that is engaged with using it for certification, with very few exceptions LEED is irrelevant to the needs and interests of the retail sector.*

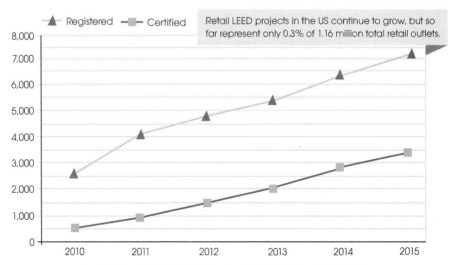

Retail LEED projects in the US continue to grow, but so far represent only 0.3% of 1.16 million total retail outlets.

FIGURE 8.5. Growth of LEED US Retail Projects, 2010–2015

Through 2014, LEED awarded 3,474 retail certifications globally (81 percent in the United States), distributed as follows:

- New Construction = 49 percent (1,688 projects)
- Existing Buildings = 14 percent (473 projects)
- Commercial Interiors = 37 percent (1,313 projects).[23]

Considering that Starbucks alone operates 21,000 stores worldwide (12,000 in US stores),[24] this total represents a very small percentage of retail stores. LEED developed its "volume certification program" specifically for retail, so a company could certify a typical store prototype, for example, in terms of materials use, energy use, water use, etc., and then just evaluate site characteristics to secure a rating. This program has existed in various forms since at least 2007. However, only a handful of retail chains use it.

Figure 8.6 shows retail project numbers (stores and tenant build-outs) choosing to use LEED during the past six years. *Here again, the bottom line is stark: LEED is not used to any significant extent by the retail market.*

The reason is plain: there is not much connection between consumer and retailer when it comes to LEED certification. Trying to make LEED a brand recognized by consumers (and also by store employees) is beyond USGBC's resources. Because of this, compa-

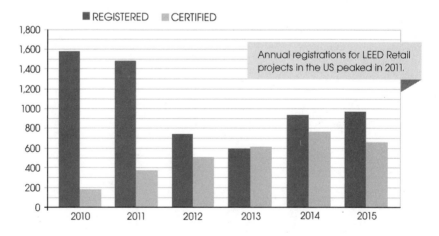

FIGURE 8.6. US LEED Retail Projects, 2010–2015

nies don't see an incremental gain in sales from spending money to certify a store. There may be other reasons, but sales growth is the most powerful motivator for retailers and LEED certification doesn't provide it.

Healthcare

Healthcare is a huge industry, accounting for 17.4 percent of US GDP.[25] LEED simply hasn't made any significant inroads into this market. Table 8.1 shows about 10,000 inpatient US healthcare buildings, totaling 2.4 billion sq. ft. and 147,000 outpatient buildings comprising 1.8 billion sq. ft. About 30 percent of LEED registered projects are over 100,000 sq. ft., typically a new hospital or large hospital wing.

Figure 8.7 shows US LEED Healthcare projects since 2010. One can see that registered projects range between 150 and 200 annually since 2012 (but only 136 in 2015), but that only 104 projects were certified in 2014 (and 96 in 2015).[26] About 81 percent were for new construction, 17 percent for tenant improvements or remodels (LEED-CI) and only 2 percent for certifying existing building operations. At 11.3 million sq. ft. certified area, 2014 LEED healthcare certifications amounted to less than 0.3 percent of the total healthcare building area.

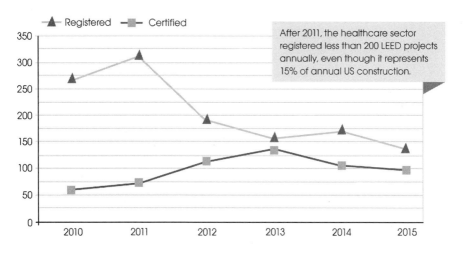

FIGURE 8.7. LEED US Healthcare Projects, 2010–2015

It's hard to assign a market share to LEED in the current health-care market, because many projects are small enough to be outpatient facilities (e.g., clinics) and not hospitals or inpatient facilities. But LEED's healthcare registration and certification numbers are strikingly small for such a huge market, and there is no indication that real growth will occur in this vitally important building segment.

Existing Buildings

We know that if LEED is going to have a major impact on 2030 goals for reducing carbon emissions, it must penetrate the existing build-ing market. To date, LEED's record in that market segment is quite dismal. While LEED-EB/-EBOM came on the market in 2004 and was significantly improved with LEED 2009, in recent years market uptake has diminished dramatically.

After a major increase in certifications in 2009–2011 in the Great Recession's aftermath, as shown in Figure 8.8, during a time when new construction activity declined dramatically, by 2014 interest in LEED-EBOM certification fell dramatically to less than half the 2009–2010 level. In 2014 and 2015, LEED-EBOM registered fewer

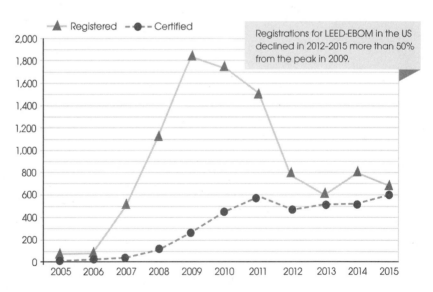

FIGURE 8.8. LEED for Existing Buildings Projects, 2005–2015

than 800 projects each year and certified fewer than 600 projects, each year totaling less than 200 million sq. ft., i.e., fewer than 0.01 percent of buildings or 0.24 percent of US building area (5.5 million US buildings; 85 billion sq. ft. of space).

The conclusion: in its present form, LEED-EBOM cannot help at all to reach our 2030 goals for reducing carbon emissions. Why do we keep pretending that LEED is making a difference in greening existing buildings? Why not abandon it in favor of something that building owners and operators will *want* to use (because it makes business sense)?

Commercial Interiors

LEED use continues to be steady with commercial interiors projects, including tenant improvements in offices, retail stores and similar activities. In this marketplace, the average project size is not large, about 30,000 sq. ft. Figure 8.9 shows the number of registered and certified commercial interiors projects, including retail. It's easy to see that this market is not growing, having leveled off at about 1,000 project registrations each year since 2010 and less than 900 certifications each year.

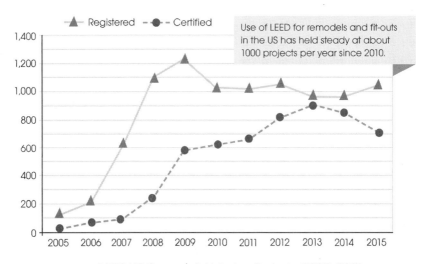

FIGURE 8.9. LEED US Commercial Interiors Projects, 2005–2015

Residential

About 884,000 privately owned US residential units were built in 2014, about 70 percent single-family.[27] LEED certified fewer than 2,000 single-family homes in 2014 representing less than one percent of such homes. The implication is stark: we're falling farther behind *each year* in building low-carbon homes, if more than 99 percent built each year don't carry a green label. Once a home is built, it is unlikely to be renovated for at least ten years, and probably not even then, so energy performance is locked in for decades, and is mainly determined by state energy codes at the time of construction.

Single-family (detached) homes and multifamily (attached) residences, both rental and owned, make up the residential market for green certification. There are an estimated 133 million existing US housing units. LEED reported[28] the following data for 53,554 *total* home certifications (since the program began in 2006) as of June 2014:

- LEED-Certified Single Family Units: 12,955
- LEED-Certified Multifamily Units: 40,599

In July 2015, LEED reported nearly 81,000 total certifications, but did not break out the total between single-family and multifamily. However, of nearly 30,000 additional units certified during that 12-month period, a good estimate would be that 80 percent were multifamily units.[29]

In 2013 LEED certified a total of 18,000 residential units, of which about 75 percent (13,500) were multifamily units. In that year, less than one percent of new single-family homes were certified to LEED standards and less than 3.5 percent of new multifamily units were built to LEED standards.

However, certification numbers may not represent the underlying reality. The LEED for Homes Multifamily program first became available in 2011. Since most LEED-registered multifamily projects eventually were certified, 2013 certification numbers show only how well that program did in its first full year of operation (since most multifamily residential construction projects take 1–2 years to move from registration to certification); in that respect, a 3.5 percent market share looks good.

During the 2012 to 2014 period, interest in the program increased. For example, in 2014 more than 45,000 units registered in the program.[30] In 2014 the US housing market produced about 360,000 multifamily units, and the approximately 40,000 units registered under LEED Homes represented about 11 percent market share.

It's better to look at LEED registrations and certifications together, especially to gauge the growth and overall impact of this relatively new program. There's no doubt that single-family home certifications lag far behind, but the impact that LEED certification has had on the multifamily market appears to be growing.[31]

Kelsey Mullen headed the LEED for Homes Multifamily Program for several years until the end of 2014. He says,

> There is more competition for rating systems for residential projects than there is for commercial projects. For commercial projects, it's LEED. For LEED for Homes, that's not at all the case. When LEED for Homes came on the market, there were already 80 local and regional residential programs and other national programs. Residential developers have to consider which program to use, whereas commercial developers do not because there is basically only LEED. The competition is significant. The delivery models are similar with providers [consultants] and raters [certifiers]. Most providers who serve LEED also serve NAHB and other local programs.[32]

Nonetheless, Mullen built a very successful program that is now certifying about 80 percent of all LEED residential projects. He did this by altering the delivery model so providers could perform most work, keeping third-party review to a minimum. Mullen says,

> LEED for Homes Multifamily had a 95 percent success rate [in getting certified] because the individual green raters who

were part of the certification process were involved through-out the process. This resulted in more successful submittals.[33] The reasons for doing LEED for single-family versus multi-family projects are much different. Single-family builders are trying to sell homes ASAP. Most multifamily owners will own a property for a number of years. They may rent a prop-erty until it stabilizes, then sell it. Or with an affordable devel-opment, they may own a building for 15 years. In this respect, multifamily developers are more like commercial developers [and have the same business reasons for using LEED].[34]

One might expect continued use of LEED for Homes in the future multifamily market, if LEED continues to use this very different delivery model, wherein the consultants also provide most data for certification at very low cost per unit. This is pretty much the same delivery model used by BREEAM and Green Globes. The residential market is extremely cost-sensitive, and only a very low-cost system will work.[35]

Summary

It's hard to summarize so much information, but the overall conclu-sion is straightforward: cumulatively, total LEED project certifica-tions in LEED's most developed market, the United States, hover at less (often far less) than one percent of the building stock and less (often far less) than four percent of the building area in almost all market segments.

Judged by this criterion alone—market penetration—*LEED has significantly failed to transform the market* for new green buildings (ex-cept large commercial office and corporate buildings) and is making almost *no headway* at all in the market for greening existing buildings, the most important carbon-cutting task for a certification system. *No amount of hype, PR or "movement" rhetoric can obscure this situation.* Therefore, our task is to understand more precisely why this is the case and what we can (and should) do to rescue this failing experi-ment in building sustainability.

CHAPTER 9

Forensics: Why Green Building Has Hit the Wall

When you have eliminated the impossible—whatever remains, however improbable, must be the truth.

Sherlock Holmes[1]

So far, we've discussed LEED's substantial merits and considerable drawbacks and we've documented that the US green building movement, exemplified by LEED, has stopped growing and is on track to *reduce* its market share over the next five years. Before we get to suggestions about what should change to push green building certification back onto a growth path, we need to understand *why* this has happened.

Let's apply some elementary forensic principles: *first analyze the situation, then make a diagnosis, assign causes to failures and only then offer a prescription for fixing the problem in the future.* Since Green Globes and Living Building Challenge together have only limited market share, certainly less than five percent of the current US green building certification market, we'll focus the analysis primarily on LEED.

What has gone wrong? LEED's success in high-end commercial real estate in "top-tier" markets, for top global corporations and also for some top-level universities and colleges, is certainly important, but these segments do not represent the overall market. For all the rest, LEED has become increasingly irrelevant. Why? There are some obvious candidate reasons, shown in Figure 9.1 and discussed below.

1. It's too hard to get certified.
2. Documentation is cumbersome.

3. Arbitrary and capricious rulings.
4. It costs too much.
5. It takes too long.
6. There's no longer much PR value.
7. Idealists designed LEED; realists rule the market.
8. LEED is too rigid.
9. Certified vs. "Certifiable."

There is perhaps another important and overlooked perspective, the marketer's: there just may not be a very large market for green building certification at LEED's current price point and value proposition. In that case, the *potentially addressable certification market* may be saturated and it's time to look at expanding the market by lowering

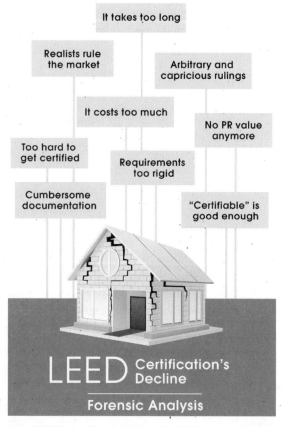

FIGURE 9.1. LEED Certification's Decline: Nine Possible Reasons

costs dramatically and simplifying the entire certification process. However, first we need to understand what's gone wrong.

What has gone wrong? A short answer would be "all of the above," with each cause in its own way inhibiting building owners from pursuing LEED certification, especially for existing buildings; taken as a whole, they work against any growth in system use and may actually lead to its decline over the next five years.

Fundamentally, in trying to please everyone while keeping the basic structure of the system in place, during the past five years LEED just got bloated: 22 different rating systems,[2] credit interpretations by the thousands, addenda by the hundreds and alternative credits by the hundreds all vetted in a "credit library," which requires an army of LEED specialists just to give reasonable guidance to projects.

GBCI and LEED have never proposed altering the delivery model to get away from using their own third-party review teams and to use instead the approach taken by BREEAM and Green Globes (as well as quality management and certification systems worldwide), which is to train, certify and monitor third-party assessors to deliver the certification. A few years ago, LEED and GBCI took one step to alter the delivery model, when they created a role for a few "Proven (or Preferred) Providers" who have relatively lesser oversight.

Talk to any experienced LEED consultant and you'll get an earful about high costs, long delays, inconsistent rulings and other problems that are endemic to the way in which GBCI operates LEED. Probe further and you'll get answers like "it's the best we can do, given the system."

Let's examine in greater detail some likely causes for the drop in interest in and use of LEED.

It's Too Hard to Get Certified

Is it possible that LEED's requirements are often unreasonable and not in line with contemporary design and construction practice? Perhaps, but the reality is that a new construction project that aims at LEED certification will find it pretty easy to get certified, *if* it meets all prerequisites and remembers to document specific criteria at the appropriate time. However, unusual project types may find that getting an energy model accepted could be the biggest obstacle, since meeting the minimum energy performance prerequisite depends on using a specific ASHRAE modeling protocol. In some instances, even low-energy-using projects may have a hard time meeting the energy prerequisite. An example is projects in regions that don't use any air-conditioning, such as the Caribbean or Hawaii.

Forensic Judgment: LEED won't work for all projects, but for the most part, especially in new construction, it's *not that* hard to get a project certified.

Documentation Is Cumbersome

Everyone who has worked with LEED has had issues with documentation requirements. As we'll show in later chapters, there are ways to reduce this burden by using technology to take information directly from construction drawings or building operations software. But a better approach would be: How little information can you review to confirm that someone has complied with a credit requirement and still keep the system "legitimate"?

Art Gensler founded the world's largest (and eponymous) architectural firm and is an immensely practical person. USGBC founder David Gottfried quotes Gensler's view:

> I recall that meeting with Art Gensler when he reemphasized the 80/20 rule. "We can streamline the process and still achieve most of the results we want," he said. He felt we'd gone way too far toward demanding perfection in our documentation and certification review process. If someone really wanted to cheat that badly, we should do our best to catch them with simple flags and screens, just as the IRS did with taxpayers. And he had another good point: Why did we require so much documentation as part of the LEED paperwork submission? It made me think of my dad's motto: *Do not confuse effort with results.* Often the documentation had little to do with the ultimate and actual results.[3]

This brings up an interesting point: If the US Internal Revenue Service can process 240 million tax returns (including 145 million individual returns) each year and keep most people honest by using software screening to examine returns and only auditing less than one percent of returns,[4] why does LEED need to review *all* submittals in detail using individual professionals, to handle 4,000 certifications?

Forensic Judgment: LEED needs to comprehensively overhaul its documentation requirements and decide how to take all the information it needs directly from electronic design documents and already available operating data. Don't you think that design software could calculate if 90 percent of workspaces had views to the outdoors directly from floor-by-floor layouts and provide that information without requiring the drawing of an office layout?

Arbitrary and Capricious Rulings

All experienced LEED consultants have their "war stories" about arbitrary and nonsensical rulings from LEED reviewers that require them to take additional measures to meet criteria required for certification, driving up costs and extending the time required to get a decision on certification. In addition, such rulings make building owners and project teams question why it's so difficult for their consultants to understand the LEED system.

Often review teams hired by GBCI create additional problems, with excessive rigor, arbitrary rulings and changing requirements for certification. For many consultants, even those with years of LEED certification experience, this can be off-putting. Often, review teams' requirements may seem academic, with more required than demanded by the LEED credit language or intent. Things can also change as reviewers realize that they need to tighten up how something's written. Eventually these decisions may appear as an addendum, but for consultants it's always a moving target, and some experience that they never quite have it all figured out.

For example, one LEED for Schools prerequisite (and only for that rating system) is "minimum acoustical performance" (IEQ

Prerequisite 3). Yet an experienced (and very practical) LEED user argues that acoustic design has nothing to do with sustainable design:

> LEED for Schools has a minimum baseline requirement for acoustic separation of classrooms. But [in this project] they were designing a three-walled classroom based on educational research about how that setup is more conducive to collaboration. [This project] was one of the most sustainable buildings I've worked on. However, certification almost didn't happen because the school wouldn't qualify for LEED, as it didn't meet the baseline requirement for that acoustic separation. USGBC was overreaching in that case because *acoustic separation has nothing to do with sustainability*. In the end, the project was able to get that decision overruled and get the building certified.[5]

By contrast, in Green Globes, the assessor makes the rulings on criteria eligibility directly, subject only to final review by the Green Building Initiative. In that system, a project team works directly with a peer professional who will pass judgment on its achievements. It can ask questions and get straight answers to rely on. This is not the case with LEED, as review teams have the final say, and for the most part they are not accessible during the design process.

In BREEAM, once assessors are trained and certified, they make primary judgments about project achievements, subject to final oversight by the certifying organization, BRE. A typical BREEAM certification is issued within three to four months after submittal of project information.[6] LEED for Homes Multifamily program works this way also, resulting in an improved customer experience and a high percentage of certified projects.[7]

Forensic Judgment: There is indeed much arbitrariness in LEED, but with a good consultant and enough project money to pay for additional work by consultants, most projects can get certified—although not always at the certification level desired.

It Costs Too Much

Every project has a budget and, if LEED certification is a goal, then it must be included in the budget. Still, many project owners question why it costs "so much" to get certified. A detailed analysis shows why certification's "soft" costs are hard to reduce, given that most project teams must hire consultants (or pay for in-house professional staff to do the job) to get through the credit interpretation maze. In this way, LEED imposes a "tax" on each project (see Figure 10.2).

The term "soft costs" is misleading: It refers to costs not directly involved with site preparation or construction, but includes fees for architects, engineers and consultants, including LEED consultants. When it comes time to write the check for LEED certification activities, however, these costs are just as hard as all the others.

In addition, there are "hard" costs to meet LEED prerequisites and credit criteria, through more advanced energy-efficiency measures, for example. LEED advocates have published several studies over the past decade that show one can design a LEED Gold project at the same capital cost as a conventional project, but this isn't always the case.

A 2011 review took an additional look at the cost of certifying branch bank buildings at the basic LEED Certified or Silver levels and concluded that there was a 1.5 percent to 2.0 percent *increase* in total building cost.[8] Proponents argue that energy savings alone will eventually pay for extra certification costs and that certification is necessary to ensure that a project realizes these savings.

For most projects, higher levels (e.g., Gold and Platinum) typically do cost more, but proponents argue that extra costs are justified by business benefits. This is perhaps true for commercial offices in downtown areas of large cities, but it's hard to see business benefits for most other project types, except for a greater potential for energy savings.

Here's the kicker: The same energy savings could be realized at the same construction cost using LEED methods but without the extra cost for LEED certification, so shouldn't certification costs be attributed to the LEED process?

Asking whether LEED "costs too much" raises another question: Compared to what? The real question in practice is: How much *value* is created by the *cost*? Some owners see value, while others just see another project cost to be controlled or avoided. In the next chapter, we discuss in detail the issue of value.

For existing buildings, where LEED-EBOM just assembles best operating practices into a highly complex rating system, fewer than 600 projects were certified in the United States in 2015 (about 0.01 percent of 5 million existing buildings), which suggests that most owners do not perceive enough value from a LEED label to make up for the cost involved.

Forensic Judgment: LEED's extra costs are real and significant to many projects. Often when an estimated total cost for LEED is presented, an owner decides to be "LEED Certifiable" instead.

It Takes Too Long

LEED certifications can drag on for far too long as review teams engage with a project team. The last LEED project I took through to certification took more than a year after all documentation was submitted to get a final award at the LEED Gold level. By that time the project team had moved on to other work and the certification no longer had much PR or marketing value for the owner.

In Green Globes, assessors must get the final project review done within 30 days after the required onsite meeting. Not only that, they must issue a detailed report (typically 15 to 20 pages) on all credits granted. At that point GBI has up to 45 days to review the report and award a certification. So within three months after the assessor's site visit, projects get a decision on certification. Why does LEED take so long, and how can the long review process be justified to building owners? The short answer is: there's no good reason, and there is no way the lengthy review process can be justified.

Rich Michal at Butler University has considerable experience with LEED during the past 10 years. He says,

One of the challenges—the shortfall—of the LEED system is just the expense and the time it takes. It's ridiculous how long it takes. It's good that you've got those independent third-party folks reviewing everything, but there's just got to be a more user-friendly and easier way than the way it's been run in the past.[9]

USGBC and GBCI have fielded complaints about user experience during the certification process for at least ten years, but operate a system that cannot deliver a quick judgment because the people who know most about a project, the architect, engineer, LEED consultant, or person who submits project information via LEED's online portal, cannot make the final assessment.

Forensic Judgment: It's certainly hard to market an intangible product like LEED that takes so long to render a judgment. Imagine spending tens of thousands or even hundreds of thousands for a "label" and not getting an answer for six months or a year. If a customer's initial experience is the key to getting repeat customers, as we all know it is, the continuing long wait to get certified is a large negative for LEED.

There's No Longer Much PR Value

A few years ago, it was still easy to be the "first" building or facility to be rated LEED Certified or LEED Platinum and to reap some public relations benefit or marketing value from that. In 2015, it's almost impossible to make the same claim, no matter how many adjectives you put after "first," e.g., "the first private secondary school in eastern Montana to be LEED Silver certified." As a result, PR's next level has become the "most" certifications, e.g., "certified more LEED projects than any other retail organization serving Fair Trade coffee in all 50 states."

Of course there's still news value today in unique stories such as Seattle's Bullitt Center (Figure C.7) which was certified under the Living Building Challenge (and also at LEED Platinum). In 2014, it documented the projected zero-net-energy performance, and the owner is promoting the Bullitt Center as "the greenest commercial building in the world."[10]

Forensic Judgment: LEED certification's value in the United States is quite small, because after thousands of such certifications there is very little "man bites dog" news value and because, in addition, most projects do basically the same thing and use similar "green" ingredients. Thus, PR value for LEED has essentially disappeared and, with it, some of the value many building owners expect from certification.

Idealists Designed LEED; Realists Rule the Market

There is no question that the LEED system was designed primarily by idealists—a small number of architects, engineers and environmental activists with a particular vision for a sustainable future.[11] To ensure support from the environmental community, it was essential for LEED to incorporate the concerns of this community in large measure.[12]

The original LEED system was designed for new office buildings, with thinking heavily influenced by the environmental issues of the 1990s: energy efficiency, environmental impacts of development, urban habitat preservation, recycling, improving indoor air quality, smoking bans, urban mass transit, dark skies, urban heat islands, daylighting, etc. With few exceptions, users of buildings such as developers, portfolio owners, facility managers, building managers, etc., were not originally invited to create the system, because they were considered more creators of problems for building design than contributors to solutions.[13]

As some clients attach importance to higher levels of LEED certification, there can be considerable "point chasing" just to raise the score, without regard for the long-term benefits. Fulya Kocak is a LEED Fellow and sustainability leader for Clark Construction, a large US commercial contractor. She says,

> A drawback of rating systems is that they can prohibit us—
> unintentionally—from thinking "outside the box." Some-

times, clients place too great of an emphasis on the scorecard and design their projects around it, when it should be the other way around. We also get ambitious with LEED and go after points that aren't appropriate for a project just to get the points; that's just human nature. When evaluating opportunities for sustainability, it's important to consider what is in the long-term best interest of the client, the project, the environment and the surrounding community—sometimes that means looking beyond traditional rating systems.[14]

The issue at hand is not whether LEED is "right" about everything contained in the system, but *whether it is offering a product that the market wants to buy.* Mr. Market is more often right than not, even though he sometimes needs a nudge to move in the right direction; for today's LEED system, Mr. Market's verdict is mostly "No thanks."

Not everyone agrees with this viewpoint, to be sure. Architect Lance Hosey makes a powerful argument that sustainable design is "a choice of values not a choice of tools"[15] and that true sustainable design *always* engages with the "triple bottom line"—ecology, economy and social well-being; or planet, profit and people. In his words, "Technology has hijacked sustainability" and removed aesthetics and beauty as considerations. One could counter, however, by saying that green building certification is all about implementing sustainable values, but that the current leading rating system just doesn't convince very many people to document their use.

A more intelligent approach to green building certification's future would start by consulting with those who are going to use the product and then would create a product that met their requirements for addressing climate change, energy use, water shortages, waste generation and ecological purchasing. We'll discuss implementing this perspective in Chapters 14 and 15.

Forensic Judgment: Environmental advocates created LEED to meet their requirements for defining an environmentally preferable built environment, but the system did not have a serious reality check with users before market introduction and it has not substantially changed its content and structure to take building owners' feedback into account. It is little wonder that most building owners do not feel any special attachment to it or any compulsion to use the product.

LEED Is Too Rigid

Most leading architects with strong commitments to sustainability don't consider LEED to be the starting point for their work. They take a broader and more holistic approach than simply checking boxes on a LEED scorecard. A project whose total carbon footprint, including both *source* energy use and building materials' *embodied energy*, is less than 100 kWh/m^2/year (about 30,000 Btu/sq. ft./year)[16] is inherently more sustainable than almost all LEED-certified projects, but if it doesn't do things in the ways envisioned in LEED's criteria, it won't be certified.

Jiri Skopek is a Canadian architect and an "old timer" with more than 20 years' experience with green building rating systems. Along the way, in 2004 he created the Green Globes system. His perspective is that fundamentals are wrong:

> It has been a two-edged sword, similar to the marketing aspect of LEED, which brought forth the awareness, but at the same time created a large cost and big bureaucracy, and developed something that I've called the "gotcha principle." Rather than this system being an encouragement for the design team to do better, it became, "I am going to make it so damn difficult; you're going to sweat; you've got to work for this." It resembles a [certain approach to] Judeo-Christian philosophy—that you can't better yourself unless you really beat yourself over the head.[17]

LEED's long (five-year) development cycle works against innovation in sustainability. Leading-edge architects, designers, engineers and builders are unlikely to wait for the next version to bring innovations to market or even to bother with trying to get LEED to certify a new approach as meriting an "innovation" credit. As an example, LEEDv4's energy credits have such obvious disregard for "zero net energy" projects that we can expect many such projects to ignore LEED.

It's true that LEED has developed many creative ways to address innovation, through credit interpretations, addenda, etc., but it's a slow-moving dinosaur in a world of fleet-footed mammals with rapidly growing concerns about climate change.

Forensic Judgment: LEED stifles innovation owing to its elaborate structure, outdated prerequisites, long development cycle and rigid approach to credit analysis by outside review teams. Most innovative architects feel free to ignore LEED in pursuing green or ecological design, leaving certification details to consultants and contractors.

Certified vs. "Certifiable"

While LEED has definitely changed the conversation around what constitutes a green building since 2000, as discussed in Chapter 6, more owners today are willing to use LEED's elements but not to certify their projects. In a way, you could say that this is a coward's choice: If you really think these issues are important, why don't you find resources to pay for third-party certification? The extra one percent to two percent of total project cost is really within the "noise" level for most projects.

Curtis Slife, head of Phoenix-based FM Solutions, says,

It's public knowledge that higher levels of LEED Certification, such as Platinum, do require additional expenditures. Clients are now asking why we just don't make it "LEED Certifiable," rather than LEED Certified. The client avoids the complicated process and saves money. It is a difficult process to go through with the USGBC, especially if they run into an issue that's peculiar, which is common with most buildings today.[18]

It's also true that this approach is less rigorous, requires less record-keeping and less attention to detail than full certification. However, it does reflect the American tendency to reject authority and substitute a "DIY" (Do It Yourself) approach. Everything considered, it's still probably better that people think about issues contained in green building certification systems than ignore them.

In some states such as California, with its CalGreen building code, and in other jurisdictions that adopt the 2015 International Green Construction Code (IgCC), compliance with code requirements gets you essentially an equivalent to a basic LEED-certified building, so why pay more just to get the certificate? We'll likely see many projects opting for this approach because it adds no extra cost beyond meeting building code requirements.[19]

USGBC has seen the handwriting on the wall in this regard and has now joined with ASHRAE and the International Code Council to further "harmonize" their requirements with LEED's prerequisites and certain specific credits. However, *just meeting the building code is not a leadership approach* and does not require the same attention to detail or expose a project to third-party scrutiny the same way as full certification, so what you get is a "LEED Lite" project.

Forensic Judgment: Expect to see more projects calling themselves "LEED Registered" with no intention to apply for certification and more that simply say they're "designed to meet LEED Silver/Gold" but never even register the project. By maintaining an arcane, high-cost, slow-to-deliver-results system, LEED has brought about this "certifiable" market and reduced its future impact.

Summary

As we stated earlier, the marketer's answer is to assert the obvious: There is just not a huge market for green building certification *at LEED's current price point and value proposition*. In other words, *LEED costs too much for the value it delivers.*

LEED is doing quite well in the rather small market segment where it truly belongs: the "1 percent," i.e., high-end markets that know how to extract value to offset LEED's costs. As it is now designed, LEED costs too much for mass adoption; 15 years into its lifecycle, one has to conclude that LEED will never reach more than one to two percent of US commercial buildings and more than eight percent of US commercial building area. It will be most successful in high-end office construction, where immediate marketing and financial benefits can offset extra costs, and in other corporate and institutional markets where noneconomic (policy) drivers are more important than cost.

Green building certification has hit the wall and has stopped growing. Is the *solution* to this problem to double down on a broken system by supposedly modernizing it, streamlining it and making it marginally more "user friendly," as with LEEDv4, or should we instead be rethinking and reinventing green building certification entirely? Can these critical issues be successfully addressed by the same organization that created them, or do we need entirely new organizations and new methods to *Reinvent Green Building* and bring about "the next green building revolution?" Think again about the example of Uber and taxis. Did putting video screens in cabs really enhance the user experience or just create another annoyance for riders?

In the next chapter, we will dive deeper into LEED's cost compared with its perceived (and real) value. This will lead us toward acknowledging that the green building certification market is highly cost-sensitive and new approaches that *cut current costs 90 percent or more* will have far more market success.

CHAPTER 10

Green Building Certification Costs Too Much

*You've got to know when to hold 'em,
know when to fold 'em*

Kenny Rogers, "The Gambler" (1978)[1]

We live in a society driven mainly by economic concerns. Companies and organizations don't want to pay for something that doesn't deliver quite a bit more value than what it costs. Green buildings are no different. People want to pay only for value received, and the value of a certification is more perceptual than actual. LEED and Green Globes certifications validate sustainable achievements, to be sure, but that plaque or certificate has value only in context.

Does Green Building Cost More?

From 2004 to 2009, several studies published by experienced cost consultants showed that one could build a green building at the same cost as a conventional building, certainly at lower levels such as Certified or Silver.[2] While that may be true, the LEED certification process itself can often add considerable cost to a project, which needs to be justified to building owners or facility managers. In addition, green buildings that achieve the top two ratings (LEED Gold or Platinum, Three or Four Green Globes) often have to include more costly products, systems and equipment to accumulate enough points to secure the higher rating.

Rethinking Cost and Value

Management studies for the past 50 years have demonstrated that human decision-making is skewed—we may pay three times as much to avoid pain (risk) as to secure gain (benefit).

Another way to think about it is this: costs are *real* and *immediate*; benefits are *speculative* and *future*. To pay for a project certification, an owner has to write a check *today* for LEED fees, for extra capital costs and for consulting services to meet LEED (or Green Globes) criteria, create energy models and provide building commissioning services. Then the owner may have to pay extra for architects, engineers, landscape architects, contractors and others to prepare and upload specific documentation to meet LEED requirements and then even more to respond to requests for clarification from LEED reviewers or GBCI staff. Finally, owners not intimately familiar with and experienced in using the LEED system must often engage a green building consultant to supervise the process.

The last LEED project on which I served as a consultant (Figure 10.1), a $40 million, 250,000 sq. ft. corporate project in Arizona, paid my firm $70,000 to manage the certification process and also to conduct a one-day green design "eco-charrette." The project incurred additional costs for a leading building engineering firm to create two different energy models: one for a "conventional" (barely code-compliant) building and one for the final design (along with several intermediate design iterations). It cost additionally about $75,000 for the LEED-required building commissioning. In addition, the owner paid the project team for additional professional services to prepare documentation to meet LEED requirements. Yet the LEED Gold certification took well over a year to be delivered to the owner, negating any PR value.

An experienced LEED consultant can certainly deliver value to a project, but LEED certification has become so complex that a consultant is almost a necessity, just to navigate alongside the shoals of the certification process. Why should we accept this situation? An Energy Star label has value (and is used three times as much as LEED, as shown by the NGBAI report discussed in Chapter 6) and it can be

FIGURE 10.1. UniSource Energy Building, Tucson, AZ. This 250,000 sq. ft. corporate office building in downtown Tucson, AZ, achieved Gold certification in 2012 under LEED-NC 2009. Developed by Ryan Companies and designed by The DAVIS Experience and Swaim Associates, this project provides a modern workspace for 425 employees. Credit: K2 Creative LLC

secured by eligible buildings with very little effort beyond uploading operating data, along with a little information on building characteristics and certification of results by a licensed architect or engineer.

What I Learned as a Management Consultant

For about 12 years, during the 1980s and 1990s, I worked as a management consultant for small businesses, firms generally ranging in size from $5 million to $50 million in annual sales. All were privately owned companies. During that time I worked with more than 75 CEOs, examining operations and preparing analyses, and each week I offered detailed recommendations for improving operations — typically by cutting costs and changing management policies and procedures. I worked with manufacturers, contractors, distribution companies, service providers and similar companies.

Getting paid required showing immediate value: I came into a business on a Monday (sometimes with others on my team) and each Friday, when I collected the weekly consulting fee, I had to identify (and justify) enough benefit to the business owner not just to cover my fees, but to provide a first-year return of three times my cost *for that week*! If I couldn't make that case, I often wasn't invited back the next week to continue the assignment. Tough love, but that's the real world!

In other words, I was not offering a "payback" on investment within two years or three years, i.e., a 33 percent to 50 percent return on investment (ROI), but a 3-to-1 first-year benefit, or a 300 percent return on investment, a four-month payback. That's value, not cost!

This experience taught me that the decision-maker wants to see benefits that are *real* (to them) and *immediate*. This example shows that we need to completely rethink what green buildings (and especially certification) *should* cost. Rather than taking the (seller's) perspective that the cost is justified by the need to meet various requirements in the certification system, we should instead take the buyer's perspective: What (net) value will I get from this service or these design measures? What's in it for me? Why should I pay more (to you) to do this?

What Should Green Building Certification Cost?

When examined from this perspective, the bottom line is clear: green building certification needs to cost *far less* than it does today, and certification needs to happen *much faster*. There's no way to grow market acceptance without a radical revamping of the entire process, starting with cost and review time. A system that needs specialized consultants and very high-level professional services to document results simply won't work for the broad marketplace, as it doesn't provide real value to building owners, who, after all, must underwrite the process.

But let's now consider another aspect: LEED is essentially a "tax" on building projects, one that continually needs to be justified with positive results.

The "LEED Tax"

USGBC and others have tried to make a case that green building creates millions of new "green jobs," both in products and services.[3] In many ways, this is a specious argument: On the one hand, advocates say it costs no more to build green. In this case, how can spending no additional money create additional jobs? If building green does cost more money, as many in the industry believe and can demonstrate from experience, then additional costs act as a *disincentive* and work just as much to reduce green jobs as to create them.

It is also true that the green building revolution has led to many innovations in products and project delivery methods and that many products we use today in buildings are quite a bit better than those on the market 10 to 15 years ago. Green building supporters should rightly trumpet these achievements.

However, there is one area where LEED has to hang its head in shame: the "LEED tax." Think about it this way: In 2014, LEED certified about 3,400 US projects, about 2,000 in new construction, 850 in interior fit-outs and 550 in existing buildings. *At an estimated average project cost of $136,700 for LEED-related items, the LEED tax in 2014 was about $382 million* (Figure 10.2 and Table 10.1).

Over the next five years, at current certification rates, the LEED tax could easily cost the building industry nearly $2 *billion*, not counting any extra construction costs for LEED-required measures. Two billion dollars is a high price for the building industry to pay to implement greener buildings in a small fraction of all buildings! Given this tax, is it any wonder that LEED's market penetration is so low?

This tax *excludes* any additional capital costs for more energy-efficient products and systems; green roofs; the famed (and often little-used) bicycle storage lockers, changing rooms and showers; specially-sourced green materials, etc. Since about 40 percent of certifications are LEED Gold and Platinum projects, one can expect many to incur higher capital costs as well as to pay a higher LEED tax.

It is hard to believe, but for 15 years LEED has not tried (very hard) to reduce the cost of certification. USGBC and GBCI have not taken any serious action to cut costs dramatically, e.g., by changing the "delivery" model, despite overwhelming evidence that it was not reaching a broad market. Instead of taking away everything from the LEED system that is not essential to a green building and driving costs down relentlessly, USGBC and GBCI have piled on more complications, more credit interpretations, more addenda and more prerequisites (in LEEDv4, see Chapter 11).

TABLE 10.1. The Annual LEED Tax (US Nonresidential Projects)

Cost Element	Cost Per NC/CS Project[4] (150,000 sq. ft.)	Cost Per EBOM Project (150,000 sq. ft.)	Cost Per CI Project (50,000 sq. ft.)	Total (@3350 LEED Certified Projects/ Year)[5] $ Million
LEED Consulting Services	$40,000[6]	$80,000[7]	$30,000	$147.5
Building Commissioning	$60,000[8]	$10,000[9]	$10,000	$131.0
Energy Modeling/Audit	$15,000	$ 7,500	$ 5,000	$37.5
Other Professional Fees (includes other analyses to meet LEED criteria)	$15,000	$15,000[10]	$10,000	$45.5
GBCI Registration & Certification Fees	$6,700[11]	$ 7,200	$ 4,200	$20.6
Total	$136,700	$119,700	$59,200	$382.1

You may argue that this is a small price to pay for better buildings, and you may be right. But you should then also acknowledge that LEED's high cost acts as a disincentive for many project teams to pursue high-performance green building.

It's time for a major change in approach, one that lowers costs dramatically—it's time to reinvent green building for the "other 99 percent!"

Lessons from Other Economic Sectors

The lesson everyone learns in the building industry is that costs matter—a lot. Return on investment, "payback" and "total cost of ownership" are nice concepts for sellers, but for most buyers upfront costs matter the most. In the US economy for the past 30 years, sellers have

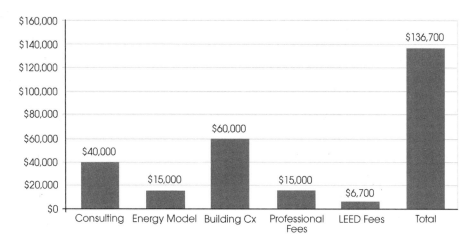

FIGURE 10.2. The LEED Tax for an Average Project

relentlessly worked to cut costs and improve product quality. So why should green building certification systems be exempted from the same logic?

Based on my experience as a management consultant, I'm advancing "Yudelson's Green Building Value Theory": anything that we ask building owners and operators to buy (such as certification) that isn't essential to their project construction or ongoing operations should be prepared to demonstrate a 3-to-1 first-year return (i.e., a four-month "payback"). Otherwise, we'll have to rely solely on individual commitments from "enlightened" owners and policy measures from government agencies, but market uptake will stay small.

The market's judgment is brutal, immediate and broad; that's why green building certification must get far less expensive and far more inclusive, to move beyond its currently well-defined but narrow (and ultimately self-defeating) niches.

Summary

We have argued that user experiences, overall social benefits and cost-effectiveness must be central elements in any green building certification system that expects to gain widespread market acceptance. For a product or service to gain traction in the market, benefits must be immediate and tangible, and they must be significantly greater than upfront costs.

By these standards, as well as by the dramatic reduction since 2010 in LEED project registrations in its major market, the United States, *the market has already rendered its verdict: "This is not for us."* Rather than take these concerns seriously, LEED and USGBC have decided to "double down" on a risky bet that an even more complex, more prescriptive LEEDv4 will entice more building owners to use it. In the next chapter, we explore LEEDv4 and give our reasons why it is likely to be a costly disappointment, one that may very well *set back* the cause of green building that it seeks to advance.

PART III

LOOKING FOR SOLUTIONS

CHAPTER 11

LEEDv4:
Can It Succeed?

*When you're up to your ass in alligators, it's hard to remember
that your original intention was to drain the swamp.*

Anonymous[1]

Is LEED version 4 (LEEDv4) a solution to cost and value issues described in previous chapters? In my view, what began in the late 1990s as a reasonably straightforward method for promoting green building design in new construction just keeps getting more complex and costly.

LEEDv4, developed to replace LEED 2009 and introduced in November 2013, was originally scheduled to become the only usable LEED system as of June 2015. After strong objections from industry about certain materials credits, including the assertion that USGBC had ignored valid scientific evidence in developing those particular credits, the effective date was postponed until October 31, 2016.[2] At that time, any new project that registers for LEED certification will have to use LEEDv4. However, projects that register under LEED 2009 before the 2016 date will be allowed to certify under the older system until June 30, 2021, giving them nearly five years to finish.[3]

Because it was developed in 2011 and 2012, LEEDv4 basically locks in those years' thinking until beyond 2021, just as LEED 2009 (LEED version 3) locks in 2007–2008 thinking until 2016 and even all the way until 2021. Today's world moves too fast for a so-called "Leadership" standard to evolve so slowly.

Some experts we interviewed expect LEEDv4 to fail to grow the market, and some think that it may even damage green building certification's future. One leading California-based consultant, with 15 years' experience working with hundreds of LEED projects, predicted, "Once USGBC completely converts over to LEEDv4, it's going to die."

What Is LEEDv4 Supposed to Do?

According to USGBC, LEEDv4 "is bolder, more specialized and designed for an improved user experience." USGBC says LEEDv4 is better because it:[4]

- Includes a focus on materials that goes beyond *how much* is used to get a better understanding of *what's in the materials* specified for buildings and the effect those components have on human health and the environment
- Takes a more *performance-based approach* to indoor environmental quality to ensure improved occupant comfort
- Brings the *benefits of smart-grid thinking* to the forefront with a credit that rewards projects for participating in demand-response programs
- Provides a clearer picture of *water efficiency* by evaluating total building water use[5]

Consultant Chris Forney agrees; he thinks LEEDv4 has created a better system.

> LEED has been quite successful in LEEDv4 in making credits more relevant. For materials credits, the current state of awareness in the building industry has advanced significantly, to where recycled content and regional materials are now commonly understood. The pivot point for market transformation has moved to where we can now work on transparency with regard to what ingredients product manufacturers put into our building materials.[6]

What's the market's verdict? As of December 31, 2015, project teams had had LEEDv4 available for two full years. If it were genuinely a better and easier to use system, as USGBC and LEED's supporters claim, one would expect significant adoption to have occurred, but the data indicate otherwise (Figure 11.1). During the first two years of availability, less than six percent of all nonresidential LEED projects registered under LEEDv4, and only 22 (total) were certified.

There is little reason to suppose, other than putting a (figurative) gun to the head of users, that most project teams will start using LEEDv4 before it is mandated, *except* for committed consultants and building owners preparing to use LEEDv4 after October 2016 by trying it out now. If it were truly easier and cheaper to use, LEEDv4 would have accrued a lot more than six percent of total project registrations during the first two years it's been available.

Indeed, if the major problem with LEED is cost and complexity compared to perceived value, it's hard to believe that making the system *more costly and more complex* will do anything but hasten the ongoing decline in LEED's use in the US building industry.

In my view, the four major benefits for LEEDv4, cited above, constitute a remarkably short list, when huge improvements are needed to get more people to use the system. Particularly missing is a much stronger, almost singular focus on reducing carbon emissions by more strongly supporting renewable energy use and zero net energy buildings.

USGBC also says, "LEEDv4 includes new market sector adaptations for data centers, warehouses and distribution centers, hospitality, existing schools, existing retail and mid-rise residential projects."[8]

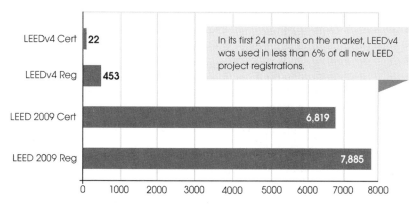

FIGURE 11.1. LEEDv4 US Projects vs. LEED 2009 Projects, 2013–2015 (25 Months)[7]

However, creating new market sector adaptations for a standard that is used by relatively few projects each year seems akin to rearranging deck chairs on the *Titanic*.

The LEED Delivery Model Is Still Wrong

One major problem is that LEEDv4 still does not change the delivery model, which one experienced LEED consultant characterized as "prove it to me,"[9] that requires a project team to submit documentation for each claimed credit to a review team and to respond to all reviewer comments, until such time as the review team accepts or denies specific credits. The same consultant pointed out, rightly, that *most owners don't see the value in this extensive review process.* Owners rightly feel that if you're smart enough to be hired to design their buildings and you're accredited by GBCI as a LEED certification specialist, you should be smart enough to document their projects and give them a final certification. So do I!

Why can't LEEDv4 adopt a new approach to documentation review and certification, one that works well in other systems, and thereby cuts time to finish (and hassle) dramatically? BREEAM provides initial review comments back to assessors within three weeks, and the entire review process takes only three to four months;[10] Green Globes certifies projects within four to six weeks after assessors conduct an onsite review and submit a written assessment report.[11]

LEEDv4—New Construction Adds to the Problem

A significant issue with LEEDv4 is the expanded use of prerequisites. Because a project must meet all prerequisites, some nonsensical in 2015 (and beyond); because many prerequisites have either become code or are being widely observed (e.g., when did you last go into a US commercial building that allowed smoking?); and because projects come in all sizes, shapes and approaches to design, LEED has had to devise many work-arounds to help projects get certified.

Let's look at prerequisites in LEED 2009 and compare them with what's in LEEDv4 (Table 11.1):

TABLE 11.1. LEED-NC: LEEDv4 Prerequisites vs. LEED 2009 Prerequisites

Prerequisite	LEED 2009	LEEDv4	Comments
Integrative Design	—	Integrative Project Planning & Delivery	Tries to standardize what is a very complex process
1. Sustainable Sites	Construction Activity Pollution Prevention	Construction Activity Pollution Prevention	This is standard industry practice
	—	Environmental Site Assessment[12]	Standard industry practice
2. Water Efficiency[13]	—	Outdoor Water Use Reduction	30%+ reduction in outdoor water use
	—	Indoor Water Use Reduction	20%+ reduction from building code
	—	Building Level Water Metering	Nonsensical; almost every building has a water meter
3. Energy and Atmosphere	Fundamental Commissioning	Fundamental Commissioning & Verification	Adds O&M plan to earlier prerequisite
	Minimum energy performance	Energy use 5% below ASHRAE 90.1-2010	Weak requirement; way out of date
	—	Building-level energy metering	Share data for 5 years; big deal!
	Refrigerant management	Refrigerant management	Unnecessary; CFCs are not used anymore
4. Materials & Resources	Storage & Collection of Recyclables	Storage & Collection of Recyclables	Good idea, but why is it a prerequisite? What about towns that co-mingle building waste and sort it at a transfer station?
	—	Construction & Demolition Waste Management Planning	This is a standard practice in construction and is code in most cities.
	—	PBT Source Reduction: Mercury[14]	Standard practice for health-care operations; why is it a prerequisite?
5. Indoor Environmental Quality	Minimum Air Quality Performance	Minimum Air Quality Performance	Has unintended consequences for more energy use in some climate zones
	Environmental Tobacco Smoke Control	Environmental Tobacco Smoke Control	Most buildings are non-smoking in today's world
	—	Minimum Acoustic Performance[15]	Can work against good design in some schools

LEEDv4 Will Likely Cost Even More

What's changed in LEEDv4? In the New Construction rating system, LEED moved from 7 prerequisites to 12 for most project types and to 14 for Schools and Healthcare. These new prerequisites, along with new credit requirements for standard credits, may add considerable cost, complexity and uncertainty to new building design (Table 11.2). It's hard to imagine that owners on tight budgets will be willing to pay for consultants' learning curves, and it's equally hard to imagine a smooth transition in credit rulings and determinations from review teams. Potential cost increases shown in Table 11.2 do not account for possible higher capital costs occasioned by LEEDv4's requirements, but apply only to professional consulting fees.

TABLE 11.2. Added Costs for New Prerequisites in LEEDv4: NC, Schools and Healthcare

Prerequisite	Cost	Comment
1. Integrative Design	$10,000+	Uncertainty about what will qualify
2. Environmental Site Assessment (Schools)	Standard analysis for any polluted site	Unnecessary paperwork for LEED
3. Outdoor water use reduction	Unknown	Requires calculations by landscape architect
4. Indoor water use reduction	Depends on building size; may be large if cooling tower is used	Requires water-efficient fixtures & practices
5. Whole building water meter	Not much	Cost for five years' reporting
6. Building commissioning & verification	$10,000–$25,000+	Requires building envelope design review & creating an O&M plan[16]
7. Building-level energy metering	Unknown	Requires annual/monthly reporting for 5 years
8. Construction & demolition waste plan	$5,000 +/-	Not yet known what planning level qualifies
9. Minimum acoustic performance (schools)	$5,000 +/-	Needs acoustic consultant; may be counterproductive to learning
10. Mercury reduction & separation	Unknown	Definitely a reporting cost

LEEDv4—"Operations and Maintenance" Doesn't Meet Building Owners' Needs

If the future of green building is going to be defined by the existing building market, then LEEDv4 for existing buildings must become LEED's major product line and be used by thousands of projects each year. However, annual LEED-EBOM US project registrations declined 63 percent from 2009 to 2015, from 1,838 projects to 674. LEED-EBOM represented only 16 percent of total 2015 LEED project registrations and shows no signs of regaining its higher usage level.[17]

USGBC could have improved and streamlined the LEEDv4 EBOM (or O+M) system to make it more usable as a management tool (thereby increasing its value) and boost its use by corporate and education facilities, and commercial building portfolio owners. It chose not to; instead, LEEDv4 *increased* prerequisites from 7 to 11, thereby making it even *more costly.*

Cost increases from prerequisites in LEEDv4-EBOM are difficult to determine, but look to be at least $25,000 *per project.* What appear to be innocuous changes really are not; for example, minimum Energy Star score requirements effectively require significant investments in energy upgrades for most facilities with an Energy Star score below 75, since to meet the energy prerequisite they are required show operating results that are 25 percent improvements (compared with previous use) in three of the past five years.

The Bottom Line—LEEDv4 Is a Bug Looking for a Windshield

Given that the 2014 Turner Construction *Green Building Market Barometer* cited in Chapter 12 indicated that most (more than two-thirds) of building industry professionals would consider using a certification system other than LEED, will LEEDv4 grow the market for green building? From our analysis, LEEDv4 appears to add cost by increasing prerequisites in an environment where first cost drives most decision-making.

The net effect is likely to be threefold:

1. It may encourage new rating systems to enter the market and compete directly with LEED, using unmet market interest in certification to deal forcefully with LEED's deficiencies.

2. LEED's overall use will decline, particularly in existing buildings, as it has done for several years, once project teams and owners realize they have good alternatives for documenting sustainable design and operations without undertaking certification's costs and risks.

3. Specialized tracking, reporting, rating and certification systems for particular market segments will take business away from GBCI and LEED; for example, the "Green + Productive Workplace" system for commercial offices (profiled in Chapter 12).

Given this likely scenario, one wonders why USGBC, as the leading organization promoting green building, does not admit the current system's shortcomings and take more forceful steps to reduce LEED's costs dramatically, change the delivery model, and commit to a new approach that would aim at getting 25 percent of US buildings certified by 2025? At the current rate, the total US building area certified is likely to remain well below seven percent by that date, a result that, if you were seriously interested in market transformation and carbon emission reductions, you would consider completely unacceptable.

With higher costs, more stringent requirements and no change to the delivery model, it's hard to escape the conclusion that LEEDv4 is a bug looking for a windshield.

CHAPTER 12

Current Alternatives Won't Solve the Problem

*I think frugality drives innovation, just like
other constraints do. One of the only ways to get out of
a tight box is to invent your way out.*

Jeff Bezos, founder and CEO of Amazon.com[1]

If LEEDv4 isn't going to accomplish greening the building stock and dramatically reducing carbon emissions from buildings, are there alternatives currently available that would do a better job?

Turner Construction's 2014 *Green Building Market Barometer*, a biennial survey of industry executives, found that 69 percent were "extremely/very likely" or "somewhat likely" to seek certifications other than LEED if constructing or renovating a building, compared with only 41 percent responding to a similar survey only two years earlier. The number "extremely or very likely" to switch increased from 17 percent to 43 percent in 2014, a 250 percent increase! This indicates latent market dissatisfaction with LEED and a potential market for other forms of certification among building owners.[2]

Let's examine currently available alternatives to LEED and decide if can they do the job that LEED has failed to do. After reading this chapter, I hope that you will agree that most current alternatives *will not work* to grow the overall market for green building certification and conclude that *we need to start with a blank sheet of paper.* We should seek input primarily from building owners, facility managers and other users, helping us to create systems that have better business value for them while still securing positive environmental results.

In my view, the major conceptual shortcoming in all green building certification approaches to date, in the United States and in other countries, is this: *programs designed by technical experts and green advocates, often in cooperation with or supported by government agencies, don't cut it in the commercial marketplace.*

Take a cue from Apple, Google, Facebook and other successful players in today's world: if you want to have a strong market for intangible products like green building certification, it's all about the *user experience.* Bruce Duyshart, a building technology expert in Australia, suggests starting with the people who will eventually use the information about a building: building managers, facility managers, maintenance personnel and even occupants or tenants, i.e., the key stakeholders, and determining what would be useful *for them* before designing any system that will address their "pain points." He also advocates strongly for using open software systems based on international standards, so a building owner can freely choose software and can change vendors to secure advantages from systems with better capabilities.[3]

We can make a short list for what constitutes success in software[4] to serve the needs of today's mobile, interconnected world:

1. User experience is paramount.
2. User interface is critical for continued use.
3. Cost-effectiveness is critical for market acceptance.

Keeping these three issues in mind, we need to add one other criterion: *any green building certification system that gains widespread acceptance should be able to demonstrate a significant reduction in total carbon emissions from buildings.*

As we argued in Chapter 11, LEED certification costs too much. For future success, certification costs must be reduced by at least 10 times and ideally by 100 times. In other words, if it now costs $100,000 to certify a LEED-NC project, ideally the cost would be $10,000 (Green Globes certification cost) or even as low as $1,000. For LEED-EBOM, where certification costs can easily exceed $150,000, costs should come down to $15,000, at most, or preferably to $1,500.

Such cost reductions are *simply not possible* with LEED as it now operates. In addition to calling for a wholesale revamping of the rating systems to focus more directly (and to some degree exclusively) on energy, water, waste and carbon emissions, the process by which certifications are delivered needs a complete makeover, one in which software or onsite assessors make the final certification decision and LEED just monitors their work, also with software, just as the US Internal Revenue Service (IRS) examines more than 240 million tax returns but audits less than one percent.[5]

Let's assess some available alternatives to LEED. We should note that it's likely that *none of them* offers a substantial alternative to LEED, and that the real "game changer" for green building is likely to come from "out of left field." Note Uber's challenge to taxis or that, worldwide, Airbnb now has more hotel rooms on offer each night than any established hotel chain.[6]

Current major alternatives to LEED include:

1. Energy Star
2. Green Globes
3. Living Building Challenge
4. BREEAM
5. Green + Productive Workplace
6. Property Efficiency Scorecard

Energy Star

Energy Star is a well-established program that tracks energy performance for about 40 percent of US commercial building area, or 3.7 billion sq. ft., more than US LEED-certified commercial building area.[7] Created for buildings in 1999 by the US Environmental Protection Agency, it is the most widely used program for assessing a building's *relative* energy efficiency. That's precisely the drawback: it does not deal easily with assessing a building's *absolute* energy use.

Currently, 21 types of facilities can earn the Energy Star label. Commercial buildings start the process by entering their utility bill data and building information into Portfolio Manager, EPA's free online tool for measuring and tracking energy use, water use and greenhouse gas emissions, which then assigns a 1–100 Energy Star score, based on the energy use compared with similar facilities.

Facilities that score above the 75th percentile can apply for Energy Star certification. Before facilities can earn the Energy Star, a professional engineer or registered architect must verify information in the certification application.[8]

Conclusion: Energy Star is referenced in both LEED and Green Globes for evaluating existing building energy performance. It's a useful tool for that purpose. However, because it only measures relative energy performance, i.e., relative to similar buildings' performance at a given time, it won't *push* enough buildings to get to low-carbon performance anytime soon. (Energy Star's relative approach is analogous to the old story that if a bear in the woods surprises you and your friends on a hike, you don't have to outrun the bear to survive, you just have to outrun one of your friends!)

Green Globes

As a replacement for LEED, Green Globes has a lot to offer, because it is much less costly than LEED and uses a third-party trained and certified assessor to work with the project team starting at the design midpoint in a new construction project, shortening the time to final certification to three to four months, following an onsite meeting that kicks off the final assessment process. For most projects, Green Globes is cheaper, faster and easier to use than LEED, but has some significant drawbacks preventing it from gaining much traction in the marketplace for certification:

1. **Market timing:** It didn't really hit the market in a full-fledged way until after the ANSI/GBI new building standard was adopted in 2010, so it entered the market 10 years behind LEED. Most professionals who are experienced in LEED don't want to be bothered to learn a second rating system.

2. **Professional support:** Green Globes lacks LEED's "ecosystem," especially committed professionals like LEED APs and GAs who will propose using Green Globes to their clients. At mid-2015, there were about 1,600 certified Green Globes Professionals (GGPs), compared with LEED's 200,000 accredited professionals.[9]

3. **Credibility as a rating system:** LEED partisans often fault Green Globes as too accommodating to industry, specifically forest products and plastics industries, creating a credibility problem with professionals and building owners. True or not, the perception lingers among design professionals that Green Globes is "LEED Lite," so they prefer to use the real thing.

4. **Federal preference for LEED:** While the federal government recognizes it and allows agencies to use Green Globes, this determination came only in October 2013, seven years after the General Services Administration committed to using only LEED. As is the case with other users, most government agencies don't want to be bothered to learn a second system.[10]

5. **Delivery model:** With its focus on using its own certified assessors, Green Globes' current delivery model cuts many fees for LEED consultants, so they aren't inclined to recommend a system that generates less demand for their services.[11]

6. **Small user base:** By year-end 2014, GBI had certified only about 1,000 projects vs. about 30,000 for LEED. Hence, there is little familiarity with Green Globes in many markets and among most users.

7. **Policy favors LEED:** As a result of its 10-year head start, LEED became the clear choice when cities, states, universities, etc., wanted to create incentives for green buildings. In recent years, Green Globes has earned recognition alongside LEED in 26 US states and by the federal government. However, LEED still enjoys exclusive support in states such as Washington and California, and with many larger universities and even city and county governments.

8. **Similarity to LEED:** Despite differentiation in the delivery model, Green Globes and LEED basically assess the same environmental attributes in the same way, so there are few technical reasons to prefer it to LEED.

One important distinction favors Green Globes: By sometime in 2016, it expects to have an updated national consensus standard (ANSI-approved)[12] that will provide the basis for an update to Green Globes for New Construction. Given the troubles that LEEDv4 may encounter in gaining market acceptance and support, by year-end 2016 more market participants might come to see Green Globes as a viable certification alternative. Some government agencies may prefer to use a tool that is based on a national consensus standard, but that distinction appears to be unimportant to most private sector and nonprofit users.

Conclusion: In many ways, Green Globes is an admirable green building rating system, with many features that make it superior to LEED, considering cost, delivery model and user experience. However, its inability to gain a significant foothold in the US market indicates that, in its present form, it is unlikely *by itself* to spur renewed growth in green building certification.

Living Building Challenge

We described the Living Building Challenge's basic tenets in Chapter 3. LBC introduced its most recent iteration, version 3.0, in 2014, but it's unlikely to result in many new certifications.

The first obvious conclusion from the paucity of certified projects is that LBC is not a *practical* tool for judging building sustainability. This is not necessarily an argument against it; LBC was created to represent an "aspirational" system for an "ideal" green world. As the poet Robert Browning wrote: "Ah, but a man's reach should exceed his grasp, or what's a heaven for?"[13]

Second, LBC has little influence currently on evolving green building design and operations in the United States, owing to its grounding more in aspirational environmental design than in daily architectural practice.

Third, LBC effectively excludes from certification all existing buildings and most new construction. LBC version 3.0 states that its goal is for buildings to be:

> Designed and constructed to function as elegantly and efficiently as a flower…that generates all of its own energy with renewable resources, captures and treats all of its water, and that operates efficiently and for maximum beauty."[14]

By this definition, most buildings can *never* be "living," since they cannot generate all their own energy with onsite renewable resources and won't exclude all combustion (e.g., using natural gas for heating or hot water).[15] Additionally, it's nonsensical in most situations to ask that a building capture and treat all its own water for reuse. And while many would agree that it's better if buildings have maximum beauty (whatever that means), anyone familiar with architectural styles over the past century knows that accepted definitions for a building's beauty can change dramatically over time.

Zero Net Energy

"Energy" is probably the easiest concept to understand in LBC and the least controversial: a building should generate onsite enough renewable energy to meet its own annual energy requirements, using only natural sunlight (and possibly wind where available). But even this formulation assumes that a building uses mostly electrical energy, typical perhaps for office buildings but not many others.

The problem is that only low-rise (and some mid-rise) buildings can effectively have enough roof area (assuming the site area outside the roof is unsuitable for generating solar energy) to meet their energy requirements with rooftop solar, even with the most efficient solar systems, building envelopes and HVAC systems.[16]

Zero Net Water

LBC makes a fundamental conceptual error in requiring zero net water: that water and energy are similar. But energy supply is global (or regional, for large solar and wind farms), whereas water supply is inherently local or (in some cases such as the US Southwest) regional. Energy can be supplied by the market in whatever quantity is needed, whereas water can only come from local natural sources or fairly short pipelines (supplemented in only a few places by desalination).

Energy is widely available whenever (and wherever) it's needed, whereas water only falls as rain or snow where and when it wants to. Solar power is fairly constant and predictable, depending on latitude and local climate (at 340 watts/m²),[17] whereas rainfall is highly variable from year to year, making it easy to design onsite solar power to meet a building's needs, but nearly impossible to do so with water.[18]

But let's address this very troubling requirement: zero net water, as stated in the LBC:

> The Living Building Challenge envisions a future whereby all developments are configured based on the carrying capacity of the site: harvesting sufficient water to meet the needs of a given population while respecting the natural hydrology of the land, the water needs of the ecosystem the site inhabits and those of its neighbors. Indeed, water can be used and purified and then used again—and the cycle repeats.[19]

In this standard, one sees most clearly evidence of American pastoral romanticism, the notion that we should all live in individual harmony with nature, even in our mega-cities and large conurbations. In addition to impracticality for most buildings, the Water Petal's requirements may in many situations well be harmful to human health,[20] may require considerable energy to make recycling systems work and may therefore actually work *against* real long-term sustainability.

Onsite Wastewater Treatment

As a trained civil engineer and professional environmental engineer, I appreciate the work of North American engineers and public authorities who have spent the past hundred years ensuring that cholera, typhoid fever and other waterborne diseases are absent from the US and Canadian water supplies and that wastewater is treated and released without (much) impact on receiving waters. In many desert areas, such as Tucson where I live, all treated wastewater is recycled to underground aquifers for eventual reuse. Trying to infiltrate it on site is nearly impossible because of impervious soil conditions.[21]

Consider now a single building with an onsite graywater and blackwater treatment system, designed to recycle *ad infinitum* (perhaps also *ad nauseum*!) all the water used in the building, supplemented only by natural rainwater captured on the building's roof (and perhaps grounds) and treated for onsite consumption by all building occupants.

Where is the *guarantee* (financial, legal and operational) that a building owner and manager will keep an onsite wastewater treatment system functioning *as designed* for the entire building life? Still the question remains: Which building owners are going to want to pay for this learning curve or for a specialized consultant, when the health of building occupants is at stake? Further, consider liability: What building owner would want to take on the risk of operating an onsite water-treatment system that could potentially sicken building occupants?

Onsite Stormwater Management

Beyond a requirement for onsite water treatment, even more from the "Zero Net Water" Petal is nonsensical: Consider the requirement for projects to confine all stormwater discharge to a building site, which is completely impractical in most high-intensity-rainfall areas. There is no way that such sites can treat and/or infiltrate that much rain falling onto a building site or an adjacent area; it *must* flow away via storm drainage.

Conclusion: Considering the Water Petal, LBC is a romantic, unscientific dream, wherein nature cooperates with humankind by providing just enough water, at just the right time, to meet a building's needs without requiring expensive cross-seasonal storage. Moreover, all building occupants cooperate in drinking "toilet to tap" site-treated stormwater and/or wastewater, having full confidence in building owners and maintenance folks to run the treatment systems effectively at all times, no matter what the cost. This seldom happens in the real world.

For an "all or nothing" system such as the LBC, its popularity, feasibility and usability must rise or fall on its weakest link, which is certainly the Water Petal. While the LBC offers some intriguing design ideas, it is the least practical green building rating system in the United States and Canada and isn't likely to have a commercial impact—ever.

BREEAM

BREEAM is the world's most widely used rating system, with more than 425,000 buildings (mostly residential) certified during the past 20 years. It is not currently well known in the United States, although there is an International version used in 75 countries outside the UK. Figure C.10 shows a new office building in The Netherlands recently certified at the (highest) BREEAM Outstanding level.

Given that our largest need is to bring existing buildings up to par with new buildings and to improve their energy and water efficiency, BRE's "BREEAM In-Use" (BIU) rating system, introduced first in the UK in 2009 and later released in 2013 for international use,[22] provides some interesting approaches. BIU recognizes and provides three assessment options for existing buildings:

1. **Asset Performance:** deals with a building's built form, construction, fixtures, fittings and installed mechanical/electrical systems.
2. **Building Management:** deals with management policies, procedures and practices related to building operations; use of energy, water and other consumables by the base building; and environmental impacts such as carbon generation and waste disposal.
3. **Occupier/Tenant Management:** deals with how management policies, procedures and practices are implemented; assesses staff engagement; and assesses key outputs such as health and well-being, public transport use in employee commuting and energy/water use by the tenant.[23]

A project can choose one, two, or all three certification options. Typically, an owner-occupied building (as most are: think schools, corporate facilities, healthcare, retail, etc.) would use all three certification options, but the choice to assess fewer building management aspects gives the system greater flexibility. The first two assessment options deal with operating a building's core and shell. Alternatively, a company that occupies, say, three floors in an office building might choose to assess or audit just their tenant space (the third option). This is hard to do in either LEED or Green Globes, but easy in BIU. In this process, the client/building owner responds to a BIU pre-assessment questionnaire, which is then verified by the assessor and submitted to BRE for a final approval and certification.

Conclusion: BREEAM's delivery model (user pre-assessment survey, assessor review and relatively quick turnaround for review comments) would make it an ideal candidate for rapidly increasing certifications in the United States and other countries, but to succeed, BRE would need to commit to the US market and most likely to partner with some other nonprofit organization that already has a built-in customer or user base.

Green + Productive Workplace

Is it possible to give a meaningful assessment for an existing tenant facility or an entire portfolio and only charge $1,500 per site? Jones Lang LaSalle (JLL) thinks so. JLL recently launched one such system, called Green + Productive Workplace (G+P). G+P does just that, with a comprehensive survey and action-oriented reports: one for each site, as well as a portfolio report that includes an executive summary for the "C Suite" and a solid portfolio strategy for senior decision makers. Figure C.14 shows how a typical G+P report considers these issues.

For several reasons, I think G+P represents a promising future direction for existing building assessment systems. First, unlike several other systems that are primarily good for assessing current conditions, G+P is not a punishing, point-chasing experience. That's because it consists of reasonable, graduated criteria that are within norms of best practices, but that also reward exceptional measures. About 50 percent of points on the green side deal with energy and carbon issues.

Second, G+P also recognizes that a facility manager knows more about his or her facility than anyone, and so it engages facility managers as the primary information source. The entire assessment is streamlined so it can be done in one sitting without requiring a consultant. All it takes to complete G+P is a half-hour for a corporate sustainability director (or similar position) to answer yes-no questions about corporate policies, disclosures and directives, along with one hour by each facility manager (for a company with multiple facilities) to complete a detailed online survey.

As the name suggests, Green + Productive Workplace addresses both sustainability considerations and occupant wellness. Its goal is to help organizations with large portfolios to green their owner-occupied spaces or leased offices while also addressing the reality that employee health and productivity gains are—financially speaking—where the big fish live.

G+P reports are succinct and rich in information. Simone Skopek, the program's co-creator, notes that data visualization on its own—no matter how appealing it looks—is not as useful as a well-presented and thoughtful analysis that contains actionable information. That's why a G+P report does more than just provide individual facility or office scores and industry comparisons. It also specifies "who needs to do what" by assigning specific responsibilities and clearly laying out expectations: for business unit managers, facility managers, the green team, HR and employees.[24]

Conclusion: Simplicity, speed, low cost, portfolio analysis, industry comparisons, believable financial implications and actionable reports—these features make G+P a practical portfolio-benchmarking approach that can help to gradually raise performance levels from weaker buildings and facilities, while recognizing and rewarding (through a certification) those that are doing exceptionally well. One corporate facilities director said:

> Green + Productive Workplace helps organizations establish how green and productive their office portfolios currently are, *compared to what they could be*. It flags the improvements that will provide best value by demonstrating a credible

order-of-magnitude of the possible financial benefits. There is no other tool like it in the market.[25] (Emphasis added)

In my view, G+P is a solid program—comprehensive and user-friendly—that could easily replace LEED for Commercial Interiors (ID+C) and LEED-EBOM for tenants, and deliver more relevant information, much faster and at a considerably lower cost than other systems.

Property Efficiency Scorecard

The International Council of Shopping Centers (ICSC) created a Property Efficiency Scorecard (PES) that incorporates the approach we recommend, namely, keep a green building tracking system focused on key performance indicators, let the users provide input into the system's creation, and make it really cheap.[26] While not a rating and certification system, the PES focuses on the shopping center (mall) operator and is organized by four resource efficiency categories, based on services landlords provide to mall visitors and tenants.

1. Energy (mostly HVAC and lighting, but including all common energy-using services such as elevators, escalators, plus onsite renewable energy generation).
2. Water (irrigation, cooling towers, restrooms and food courts).
3. Waste and Recycling.
4. Operations (including green cleaning and electric vehicle recharging stations).

Scorecard users can assess performance of their existing buildings (both malls and open-air properties), benchmark against peers, improve the performance of their portfolio and report results. This is an easy-to-use tool, aimed at a very sophisticated client base, one that already has considerable uptake in the shopping center industry. At this time, ICSC doesn't plan to make the PES into a certification system; the real aim is to help mall owners and operators control their operating costs and to provide insight into how their properties compare, one with another, and all with industry norms. About 2,000

FIGURE 12.1. Property Efficiency Scorecard. Credit: ICSC

centers have provided operating data that are used for benchmarking individual properties.

According to ICSC, there are four benefits:[27]

1. **Assessment:** Owners enter building-performance data and measure it against standard metrics, but all data are kept anonymous to other owners.
2. **Benchmarking:** Owners see how a specific property performs within a portfolio, and also compared with industry peers.
3. **Improvement:** PES generates custom property-efficiency recommendations and helps owners to track ongoing improvements.
4. **Reporting:** PES creates standardized reports, to explain performance data to investors and stakeholders inside and outside the organization.

Costs are very reasonable: $2,000 per year for the service, plus $500 per center up to 100 centers, $400 thereafter. So for an owner of 20 properties, it would cost $12,000 per year to create reports, analyze performance, etc. That's far less than the cost of one LEED-EBOM certification (and far more useful). Figure 12.1 shows what a typical PES report looks like.

Summary

There are other systems for specialized applications such as schools and homes,[28] but there is no obvious domestic competitor right now that could upset LEED's dominance in the broader US buildings market, and that would also encourage a rapid growth in green building certification, using much lower costs and a rationale focusing on carbon reduction as a primary driver. Therefore, the remainder of this book delves into opportunities for new rating and certification programs that would drive green building more decisively into both new construction projects (beyond commercial/corporate real estate) and existing building operations.

CHAPTER 13

Is Certification
Really Necessary?

Nothing is as powerful as an idea whose time has come.

Victor Hugo[1]

There is no question green building as a concept has had a powerful attraction for many people, not just building industry professionals; otherwise, how would it be so intuitively obvious to so many ordinary citizens and so easy to explain to people in government? But the truth must be told: If all green building certification systems now on the market are destined to reach less than 5–10 percent of the US building area and less than 1–2 percent of the buildings, do we really need them at all?

Consider this simple equation:

$$25\% \times 4\% = (0.4) \times 5\% \times 50\%$$

In other words, a 25 percent improvement in energy use (about what LEED buildings are getting on average) in 4 percent of the building stock (about where LEED currently reaches) is *less than half* (only 40 percent) of what we could get with a modest 5 percent improvement in 50 percent of the building stock (or a still-modest 10 percent improvement in 25 percent of the building stock).

If our overriding concern must be cutting carbon emissions, and everything else that doesn't directly deal with climate change and its related effects is just "noise," or "nice to have" but not "must have" building attributes, why shouldn't we figure out how to get half the existing building stock to become somewhat more efficient, instead of worrying about making a relative handful of high-end buildings "really green"?

In this respect, we must ask: Has LEED become *irrelevant* to achieving building sustainability, especially with respect to cutting global carbon emissions more rapidly? If so, what new approach (or approaches) would prove more relevant and create a faster response to the carbon issue? Let's take a look at several approaches that are already used to reduce carbon emissions from buildings.

Reducing Energy Use in Buildings

There are four basic approaches that can successfully reduce building energy use, well documented in the extensive literature on water and energy conservation.[2] These four paths are regulation, incentives, policy and education (Figure 13.1).

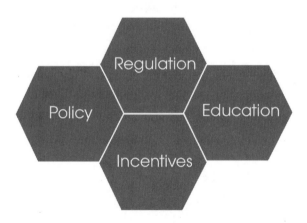

FIGURE 13.1. Four Key Approaches for Reducing Building Energy Use[3]

Regulation

From many perspectives, the easiest way to promote energy efficient and green buildings is just to require developers and building owners to build to a national standard such as LEED or Green Globes, for all buildings above a certain size. The devil is always in the details: What certification level? How to guarantee that a promise to certify will be honored? What additional costs does this impose on a developer or owner? (As we've shown in Chapter 11, these can be considerable.) What happens if the project fails to certify despite a "good faith" effort by the owner or developer? What business does a city, county, or state have adopting as a "code" the changing systems and requirements offered by private non-profits (especially those that don't use accepted consensus standards-setting processes)?[4] These are not insignificant questions, but they serve to indicate why local requirements, such as those requiring government and/or private buildings to become LEED Silver certified, don't always work well for design and construction projects and hardly work at all for existing buildings.

Building Codes. Considering mandates and regulations, there are easier ways to go, and California's "CalGreen" building code, in force since 2014, illustrates one approach. CalGreen applies to all new construction and any renovations above 1,000 sq. ft. or $200,000 in value. CalGreen requires any new building to deal with and implement best practices in such design areas as stormwater management, water conservation, energy efficiency, building commissioning and construction waste recycling.[5]

CalGreen effectively mandates much of what's in LEED and Green Globes, making certification far easier to justify to an owner ("We've already done 90 percent, so why not just pay a little extra and certify?"). This approach will certainly work in states such as California that already believe in mandating just about every conceivable behavior for owners and developers, but it will do little for the existing building stock, at least not until it is renovated (at which point CalGreen's requirements take effect), a slow process affecting perhaps one percent of buildings each year.

The push to regulate green building by code continues, with adoption of the 2015 International Green Construction Code (IgCC), promulgated by the International Code Council (ICC). The IgCC has certain benefits over LEED, Green Globes, NGBS and other voluntary systems: it takes specifying green building measures from their hands and makes it law, so everyone has to play by the same rules.

The disadvantages are twofold: First, it is a consensus standard and inevitably lags behind best practices, since the code adoption cycle is only every three years and so in some ways may even inhibit using better ideas. Second, it relies on local jurisdictions to adopt the code, an effort that can take years and involve considerable discussion with local building officials.

Incentives

Incentives are another favorite tool used by government to influence new construction. Incentives for green building come in several flavors:

1. Accelerated processing for building permits and faster entitlement approvals.
2. Increased density bonuses for developers.
3. Direct financial incentives, such as tax deductions, property tax breaks and tax credits, for efficiency investments.
4. Publicly financed loans for efficiency upgrades, using PACE programs.[6]

Incentives work ("money talks, everything else walks") *if* they influence business decisions. For example, Chicago provides accelerated permit processing for projects committing to achieving LEED or Green Globes certification.[7] If your project is time-sensitive, then it might benefit you to take this approach. However, only tax or financial incentives (or laws such as CalGreen) work consistently to encourage renovations that reduce energy use.

Policy

Local government can make it a priority to support energy efficiency goals and can do so in many ways.[8] A promising approach is to develop 2030 Districts that push local buildings to meet the 2030 Challenge.

2030 Districts. To reach 2030 goals calling for 50 percent reduction in median energy use from existing buildings and zero net energy use from new buildings by that year, these districts provide a business model for urban sustainability through collaboration, leveraged financing and shared resources.[9] Ten North American cities had 2030 Districts at mid-2015, and the movement is gaining popularity.[10] Seattle is the leading 2030 District, with participation by mid-2015 from owners of 45 million sq. ft. of commercial space.

There are many other options open to local government, but perhaps the most powerful tool is *information*. Just make information about a building's energy performance available to an informed public, so the thinking goes, and people will start to make better decisions about leasing space in energy-inefficient buildings. After all, for most commercial building tenants, who pay prorated costs based on building energy use, or for energy use in their own space (if it's separately metered), energy is often the largest uncontrollable cost for building operations.

Education and Transparency

If the goal is to gain dramatic energy consumption reductions from existing buildings, in our social media and transparency era, education is an effective tool when accompanied by public disclosure. Many cities are enacting local ordinances that require owners of commercial properties with more than 10,000 sq. ft. to disclose actual energy use in a public database at least annually, allowing them to benchmark performance against other similar buildings. Figure C.8 shows US benchmarking and transparency policies adopted by mid-2015.[11]

Education and information are important for another reason: *buildings don't use energy, people do.*[12] Buildings exist for people to work, play, learn, live and recreate in; a building's energy use is related to a building's physical characteristics and HVAC, lighting and power systems, but also to occupant actions. As building envelopes and HVAC systems become more efficient, the direct contribution of lighting and plug loads to total use becomes more important, reaching 30 to 50 percent in newer, more efficient buildings. People's behavior most directly affects lighting and plug loads, so we should expect education programs aimed at changing people's habits to reduce energy use.[13]

Ultimately the Customer Decides

Andrew Burr, a visiting scientist with the US Department of Energy, Building Technologies Office, is someone who has been involved for a decade at the "field level," trying to make improvements in building energy efficiency. He says,

> The key to progress for high-performing buildings is consumer demand and market competition. Real estate supply will become "greener" when commercial property and housing markets compel building owners, developers and builders to deliver a "greener" product. This is beginning to occur in some major real estate markets, but overall US market demand for sustainable homes and buildings is still quite weak. Government programs, regulations and incentives and rebates have a role in expanding the market, but ultimately it comes down to the consumer—the renters and buyers of property whose dollars influence how real estate evolves.[14]

Burr's argument cuts to the core of the green building conundrum outlined here: What do you do if the market doesn't respond as you expected? Do you ask government to help out by mandating the use of your rating system, or do you go back and re-tool the system so the market will *demand* it?

Summary

LEED became successful by selling its label as the "only" way, or certainly the "best" way, to measure a building's green-ness, to be recognized as a "leader." For other "wannabes" such as Green Globes or Living Building Challenge, it's become almost impossible to challenge LEED in commercial real estate and in the institutional market, with high-profile government projects, office buildings for large corporations and university projects.

Additionally, third-party certification is still required by those who have to report their actions to outside stakeholders, such as institutional real estate investors, lenders,[15] public stockholders, politicians, employees, school boards and university leaders.

Beyond energy, green building certification is still useful to account for other important environmental and health criteria, such as water use; materials and resource use; buildings' impacts on local ecosystems; and human health, productivity and well-being. But it's perfectly understandable that if we want all buildings to contribute to reducing global carbon emissions, then *we should insist that green building certifications do a better job in reaching a broader marketplace and that they document significantly better energy reduction performance in its certified buildings.*

CHAPTER 14

Focus on Carbon and Leverage New Technology

Begin with the end in mind.

Stephen R. Covey, 7 *Habits of Highly Effective People*[1]

Addressing climate change by reducing greenhouse gas emissions stemming from human activity has become this decade's defining issue. When one observes the Secretary General of the United Nations, the President of the United States and the head of the Roman Catholic Church all addressing climate change in 2015, it signals that this is a hugely important issue.[2] For this reason, I believe that *reducing carbon emissions from building operations (and from building products used in making and renovating buildings) should be the primary, overriding task of any green building rating system.*

Begin with the End in Mind—Architecture 2030

It's appropriate to ask: What is the end we have in mind? The Architecture 2030 program has the clearest "end game," envisioning that all new buildings will be zero net energy/carbon by 2030, less than 15 years from now, and that all existing buildings will use 50 percent less energy than a 2005 baseline.[3]

Architecture 2030, a nonprofit founded by architect Edward Mazria, sets the standard for cutting energy use and carbon emissions. Since the most essential thing in green building is to reduce energy use to considerably lower levels, the key is to work on existing buildings. More than 80 percent of buildings that will be consuming energy 15 years from now are already built.

Architecture 2030's goals are in alignment with what's required for mitigating US carbon emissions. Shown in Figure C.16, the "2030 Challenge" contains these goals: [4]

- From 2015 to 2019, all new buildings, developments and major renovations shall be designed to meet a fossil fuel, GHG-emitting, energy-consumption performance standard of 70 percent below the 2005 regional (or country) average/median for that building type.
- At a minimum, an equal amount of existing building area shall be renovated annually during these five years to meet a fossil fuel, GHG-emitting, energy consumption performance standard of 70 percent below the 2005 regional (or country) average/median for each building type.
- The fossil-fuel reduction standard for all new buildings and major renovations will be increased to:
  - 80 percent in 2020
  - 90 percent in 2025
  - Carbon neutral in 2030 (using no fossil-fuel-based, GHG-emitting energy to operate).

Begin with the End in Mind—
Zero Net Energy Buildings

Zero Net Energy buildings (ZNEB) are a megatrend. We are likely to see more buildings achieving ZNEB status each year, first by the dozens and soon by the hundreds, as the cost for solar continues to decrease and as more building owners see the benefit of showing their commitment to reducing carbon emissions to zero. How many US buildings could become zero net energy? Estimates range widely, from very few to as many as 70 percent, depending on the definition of "zero net." In 2007, the National Renewable Energy Laboratory developed the first widely accepted definitions for categorizing types of zero net energy buildings:[5]

A. A *footprint* renewables ZNEB. All renewable energy systems, particularly solar, are sited *within the building footprint*, mostly on a building's roof.

B. A *site* renewables ZNEB. All renewable systems, particularly solar but perhaps including wind turbines, are *located anywhere on a building site,* but not necessarily within the building footprint. This could include solar PV used for shading parking spaces, free-standing arrays, solar on top of other buildings on site, etc.

C. An *imported* renewables ZNEB. This could include biomass-fueled boilers using wood pellets, ethanol, biodiesel, etc., imported to the site.

D. An *off-site purchased* renewables ZNEB. In this scenario, a project commits to purchase enough renewable energy, from an audited supplier, to make up its annual onsite energy use. (This is of course the weakest form of ZNEB!)

In 2015, the US Department of Energy released an updated consensus definition for a zero net energy buildings as: "An energy-efficient building where, on a source energy basis, the actual annual delivered energy is less than or equal to the on-site renewable exported energy."[6] The definition can also be extended to an entire campus (e.g., school or university), building portfolio, or community.[7]

ZNEBs require a building to be designed and operated as efficiently as possible, before adding the renewables contribution. For new buildings, I have documented that a reasonable target for energy use, independent of contribution from onsite renewables, should be below 35 kBtu/sq. ft./year (111 kWh/m²/year).[8] Most well designed and operated new office (and similar) buildings can meet this target without significant cost increases.

A noteworthy example is the J. Craig Venter Institute in La Jolla, CA, near San Diego, completed in 2012. Designed by ZGF Architects LLP and Integral Group engineers, this 45,000-sq. ft. research laboratory achieves zero net energy with an energy-efficient design that features a 73 percent reduction in delivered energy use compared with a conventional lab design (Figure C.9).

Zero net is a great concept and a worthy goal for all new buildings! Missing right now is a legitimate ZNEB certification program that doesn't enmesh itself with other issues, as LBC does.

Zero Net Energy Retrofits

Retrofitting buildings to be zero net energy might be easier than you think. One favorite example comes from Singapore, with the Zero Energy Building (ZEB) at the Building and Construction Academy campus. Imagine trying to take a three-story conventional classroom/office building and making it zero net energy—that's exactly what Singapore's government did, completing an $8 million retrofit in 2009. Now imagine trying to do this in a cloudy tropical climate, almost on the equator, where temperature/humidity are 90/90 almost daily, and you'll get a picture how challenging that is. Employing many clever strategies for daylighting and natural ventilation, the building consumes only 42 kWh/m²/year (about 13 Btu/sq. ft./year). Using a 190 kW rooftop PV system, this ZEB became a net positive energy contributor beginning with its completion in September 2009. The 48,000 sq. ft. (4,500 m²) building serves 80 staff and about 200 visitors per day.[9] A meter in the lobby shows visitors net (negative) energy use since the building's opening day, demonstrating a positive net energy production from PV.

Another example is a recent California project in Silicon Valley that retrofitted a 1970s R&D building into a zero net energy 32,000 sq. ft. office, with higher rents paying the extra costs required to make it energy efficient, and with energy use 100 percent offset by solar power. It was also a commercial success, fully leased to a major tenant in three months at higher rents, compared with a local average of 18 months to complete leasing.[10]

Zero Net Energy Homes

While this book's major focus is the nonresidential market, US residential energy use accounts for equivalent greenhouse gas emissions and is more intractable to reduce, because homeowners are much more diverse than building owners. There is a second factor: except for Hawaii (where residential electricity costs almost $0.40/kWh and there is no natural gas), US energy is cheap and as a result there is little incentive for homes to conserve energy. With abundant gas supplies, residential energy costs may not increase for a long time, especially since many coal-fired power plants will likely be retired or converted to generate electricity with natural gas.

Reducing GHG emissions from homes requires a different approach, because economic incentives to gain relatively small energy savings are lacking. The core issue is not only greening the new home market, but also dealing with 133 million existing homes that must rely on the individual decision making by homeowners, landlords and homeowners' associations.

Fortunately, cheap capital and tax incentives show the way forward. Companies like Solar City[11] have effectively removed the need for an individual homeowner to invest $20,000 to $50,000 in a photovoltaic system to get solar energy's benefits, by packaging those installations as investments to third-party investors. As the current federal residential solar tax credit was extended in late 2015 past its end-of-2016 expiration date, to 2021, this approach, solar leasing, will likely accelerate its market penetration.[12]

Moving Beyond LEED, Green Globes and BREEAM—Cutting Carbon Emissions

In a prior chapter, we distilled the primary reason why current rating systems aren't going to work to cut carbon emissions from buildings quickly enough: they don't address climate change strongly enough. Each major green building rating system addresses so many diverse issues that it's easy for building teams and building operators to skimp on investments in energy efficiency and renewables, by aiming for easier and cheaper points using other measures.

The benefits of a major transition to green buildings were clearly spelled out by energy expert Greg Kats in his 2010 book, *Greening Our Built World*. In that work, Kats showed that a massive green building transition by 2020, resulting in a situation in which LEED certified 95 percent of new construction and 75 percent of existing buildings, could cut US carbon emissions by 14 percent in 2025 from a 2005 baseline.[13] By cutting annual energy expenditures, Kats estimated that there would be an economic gain of $650 billion in net present value. Clearly that scenario won't happen, but the potential for long-term carbon reductions through green building still exists. In 2010 Kats thought that LEED could lead this transition, but in 2015 it's clear that LEED isn't capable of making it happen, so we need to try another approach.

A 2014 World Bank report states that we need to decarbonize all developments to achieve zero net carbon emissions by 2100, and stresses that buildings with zero net carbon emissions are vital in this process:

> Bringing net emissions to zero will require efforts on four fronts: (i) decreasing the carbon intensity of global electricity production to zero by 2050, (ii) increasing the use of this low carbon electricity and switching away from fossil fuels—particularly in the transport, *building* and industry sectors, (iii) boosting *energy efficiency* and minimizing loss and waste, of food in particular, and (iv) preserving and increasing natural carbon sinks through reforestation or better soil management, for example. Improving *energy efficiency* and public transportation could increase global output by over $1.8 trillion per year all while tackling climate change.[14] (Emphasis added)

In my view, future green building rating systems should be at least 50 percent devoted to directly addressing climate change by radically cutting energy use, and 100 percent devoted to a few key performance indicators (KPIs) for green buildings: energy, water and waste, including induced carbon emissions from building materials (new construction) and purchasing practices (existing buildings) and Scope 3 carbon emissions such as employee commuting and corporate travel. That's it: Nothing more!

Ask the question in a different way: *What should our future green building rating systems be like?* If they are not going to resemble LEED or Green Globes or BREEAM, then what? What should the key features be and how should these systems work for most building developers, owners and managers?

We think such systems must incorporate *self-assessment* into the process as a key component, instead of using third-party verification for every project. This is a direct challenge to LEED's delivery model right now. In this approach, *third-party verification can use random, statistical sampling* rather than examining documentation details for each project; doing this would cut costs enormously without sacrificing overall quality.

In essence, any new green building rating system must have three key characteristics, described in detail in the next chapter:

1. **Smart:** it readily incorporates new technologies and new approaches for building design and operations.
2. **Simple (but not simplistic):** it does not get enmeshed in overly refining measures such as energy efficiency or trying to incorporate every sustainability nuance (such as urban heat island effect or urban habitat creation).
3. **Sustainable:** it deals with key sustainability issues, including energy use, water use and waste diversion, along with Scope 3 carbon emissions and ecological purchasing practices.

The key to any green building rating system that is likely to be widely adopted is that it has to mesh with the ongoing revolution in smart buildings. The first key to understanding why this is the case is to recall from Chapter 1 what it means to be a "smart" or "intelligent" building.

Intelligent Buildings

The term "intelligent buildings" incorporates the possibility that all buildings can be remotely managed using low-cost, cloud-based technologies. According to experts, "Intelligent buildings are those that leverage information technology (IT) to lower the costs of and speed the attainment of existing business goals."[15] In other words, intelligent buildings don't exist for their own sake, they are part and parcel of how a business develops a strategy, goes to market and seeks to make a profit (Figure 14.1), including conserving energy, managing operational efficiency, meeting sustainability goals, ensuring occupant comfort and generating a financial return.

In this rendering, "Visibility, control and policy are the functional foundations of *intelligent*,"[16] and dashboards are a key element in the tactical mix. Dashboards represent a clear view into how a building's core systems such as HVAC, lighting, plug loads, etc., are performing, which in turn allows us to better control these systems and make operational improvements.

We know that cloud-based data collection, analysis, visualization, fault detection and diagnostics, and comparisons among buildings in a portfolio are now readily available for all buildings with IP-addressable building meters and automation systems. Such systems can also easily be fitted even to existing buildings with analog-based building management systems. In addition, cloud-based systems can also address most buildings' requirements (even those that don't have sophisticated building management systems, but that nonetheless need to manage energy, water use, waste generation and recycling activity).

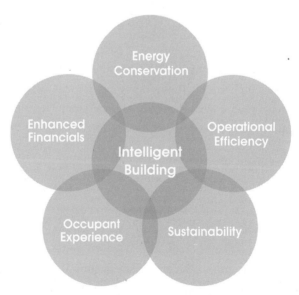

FIGURE 14.1. Intelligent Buildings Exist to Serve Business Needs

Next-gen green building rating systems should START with the idea that all data required for making them work are already available or can easily and cheaply be made available from the building operator, via vendors for energy, water, waste, etc. Data that are not readily available should NOT be included in new rating systems.

Such data categories include:

1. Scope 2 carbon emissions, i.e., electricity use: monthly, daily, hourly and even 15-minute-interval data for electricity consumption (where available)
2. Scope 1 carbon emissions (combustion via natural gas, diesel, wood pellets, etc.)
3. Water supply (indoor and outdoor)
4. Corporate purchasing ("eco-conscious" products are available from such vendors as Staples and Office Depot, for example, that have eco-labeling systems)[17]
5. Waste removal and recycling
6. Scope 3 carbon emissions such as employee commuting, corporate travel, etc.[18]

Data availability will vary by category and location, but conceptually every piece of data can be collected at least monthly and also automatically, then analyzed, normalized for weather and occupancy and displayed on dashboards accessible from any Internet-connected device, anytime, anywhere.

Vladi Shunturov, Lucid's CEO, focuses on a major US building segment, namely owner-occupied buildings, especially public and university buildings. He says:

> We're marketing to the early majority at this point, but the reason why they were early adopters in the first place was because they saw the need for this technology. They also typically are leaders—they have to lead by example and they had a sustainability practice fairly early on. So the director of sustainability, or chief sustainability officer first emerged in higher education and in cities, and it now is

pretty commonplace in a lot of enterprises. It's really because of that job function that they basically said, "Okay, now I've got to improve the efficiency of our utilities, but I don't have sufficient tools. I don't have information to make decisions; I don't have information to find out if what I'm doing is moving the needle."[19]

Unlike other building management platforms, Lucid's BuildingOS (Figure C.13) empowers a diverse set of teams to drive action from data with role-specific applications for finance, sustainability and facilities, and operations teams, through the following key solutions:[20]

1. Visibility & Reporting
2. Utility Bill Management
3. Measurement & Verification
4. Budgeting & Planning
5. Building Efficiency
6. Benchmarking
7. Occupant Engagement
8. Tenant Billing

More important than the building dashboard is the potential role that may be played by Lucid's platform, BuildingOS, in offering opportunities for many software developers to produce apps that would use Lucid's application programming interface, or API, to create customized analytical versions for specific users, much as Apple's iOS has spawned 1.4 million apps for the iPhone, iPad and other devices.[21] Such apps could include generating a green building rating score from a set of criteria.

The LEED Dynamic Plaque

USGBC introduced the LEED Dynamic Plaque (LDP) with great fanfare in 2014.[22] However, at this time it can display only five category scores from a single building. Owners can compare their building's current and past performance and can examine the building's overall performance relative to comparable structures (if they have data for other buildings).

With estimated development costs of $2.5–$4.0 million,[23] the LDP works currently only for buildings already LEED-certified. By June 2015, a scant 60 buildings had signed up to use the LDP;[24] by contrast, companies such as Lucid add thousands of buildings to their platforms each month.[25] LDP is costly as well, renting at a reported cost of $3,000 per year and requiring a three-year commitment.[26] By contrast, Lucid and Switch Automation (two leading vendors) deliver much more functionality for about one-third the cost per building, and can serve a single building as well as servicing a multiple-building portfolio.[27]

Incorporating Green Building Data on Cloud-based Platforms

Once all relevant operating data is uploaded to a cloud platform, it would then be quite easy to generate a continuous green building rating. In this way, *what a corporate CFO, building owner or facility manager might have previously thought an unnecessary expense could become an indispensable management tool.*

Lucid's BuildingOS platform (as well as platforms available from other companies) allows you to start with Energy Star and then add other data sources to get a comprehensive building performance overview, for a single building or an entire portfolio. In this way, it is not difficult to generate a green building rating for a single building, facility, or even all buildings in a portfolio, *so long as rating and certification protocols address only readily accessible data.*

What key functions does an energy dashboard or a cloud-based platform provide? (Figure 14.2)

1. All data is in one place, in (near) real time, with little effort or human error, typically in "next day" for electricity use and monthly for other services unless specifically metered with URL-address outputs.

2. Building data analysis is automated according to standard approaches, saving time and effort in analyzing data and normalizing it.

3. Data visualization is presented, not in mind-numbing Excel spreadsheets or soporific data tables, but in clear and colorful graphics. Visual display makes trends quite apparent.

4. Comparison with established or desired results is made easy by accounting for weather, occupancy, changes in use, holidays and other important variables. This allows users to easily identify "out of norm" performance.

5. Instant messaging features allow timely notification to the right person that something is out of whack; some software also allows fault detection and diagnostics for individual equipment such as a large chiller.

6. Reporting is automated, both internally to a company or agency, but also to outside bodies such as Energy Star, the Carbon Disclosure Project or the Global Reporting Initiative, freeing up valuable staff time for sustainability projects and allowing better decision-making for energy efficiency upgrades.

How Data Platforms Can Help Create
New Green Building Rating Systems

Central to this book's argument is that we need to design a new approach to green building rating systems that relies on readily accessible data. *We can radically transform green building certification by starting with a data set that every business needs (and acquires) on a regular basis for managing ongoing operations.*

This data set includes:

1. Electricity use, including onsite solar and wind power generation.
2. Gas and diesel purchases.
3. Water consumption.
4. Waste sent to landfill and percentage collected for recycling.
5. Ecological purchases as a percentage of total purchases.
6. Weather and occupancy data, for normalizing energy use, readily accessible from many global weather services.

Additional data on Scope 3 carbon emissions can come from quarterly or semiannual employee or occupant commuting surveys and on internal corporate travel reporting.

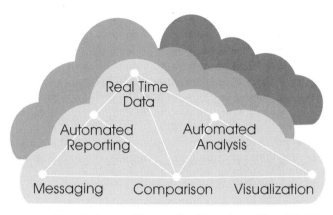

FIGURE 14.2. Key Features of Energy Dashboards on Cloud Platforms

Data that need to be collected include both *dynamic* and *static* variables. The first data type, dynamic, changes continually based on operational activity. The second data type, static, describes a building's location, occupancy type, access to transit options, landscape irrigation, habitat extent and similar location-specific data that are useful for managing a building portfolio and could be evaluated by a green building system.

If a green building rating system wants to include location information in its rating system, specific points that it awards based on such data can easily be incorporated into a platform's data output and rating scores without changing the overall approach: once a building is set up on a platform, further analysis and verification is hands-free and continuous, which cuts costs by 10× to 100× compared with using the LEED-EBOM system for assessing existing buildings.

Additionally, energy use data employed for reporting Energy Star scores can also be used to calculate absolute performance and to report that information using a scoring protocol from a Key Performance Indicator (KPI)-based green building rating system (as discussed in Chapter 15).

Cloud-based Energy Management at the Microsoft Campus

Microsoft's Redmond, WA, campus near Seattle early on adopted a very sophisticated but hands-on and real-time approach to data mining, visualization, reporting and control. Here was the Microsoft team's challenge when it started to deal with managing the entire portfolio in the cloud (Figure C.11 shows the approach taken):

- 125 buildings with 14.9 million square feet (1.38 million square meters) of office space in Redmond, WA; approximately half of Microsoft's global real estate portfolio
- 45,000 pieces of mechanical equipment needing to be maintained
- 7 major building management systems used to manage equipment
- Average *daily* consumption of 2 million kWh of energy, producing about 280,000 metric tons of carbon emissions annually

Darrell Smith is director of facilities and energy for this collection of buildings supporting nearly 49,000 employees. Over several years, he combined cloud technologies with existing systems into a smart-building operation, with an operations center staffed by ten people. Their ultimate goal: spot (and fix) operational problems (damper stuck, valve leaking, too hot/cold, too much air/not enough air) *before* they cause a programmer to have to interrupt a task to report the issue. According to Microsoft,

> Smith and his team worked for more than three years to unify an incongruent sensor network put in place during different eras (think several decades with different sensor technology from many manufacturers). The software they built strings together building sensors by the thousands to track things like heaters, air conditioners, fans and lights—harvesting billions of data points per week. That data has given the team deep insights, enabled better diagnostics and allowed for far more intelligent decision-making. A test run of the program in 13 Microsoft buildings provided staggering results—not only has Microsoft saved energy and millions in maintenance and utility costs, but the company now is hyper-aware of the way its buildings perform.[28]

Every 24 hours, Microsoft's system collects 500 million data points. Smart building analytical software then presents engineers with *prioritized lists* of misbehaving equipment. Algorithms balance out costs to fix them, evaluating both money and energy savings and considering other factors such as how much impact repairs will have on employees who work in an affected building. According to Smith:

> Today someone may be surprised, if not interested, that we're doing a smart-building deployment. I think in three to five years, you'll be scrutinized if you're not doing it. I think the market really will move that quickly. [Our goals are:] reducing energy use, optimizing asset [performance] and providing better experiences for the occupants.
>
> At our campus, I have 59,000 living sensors [i.e., people]. I don't want our employees to be sensors, first and foremost. We really value the employee experience, so what was interesting is through the use of the technology and fault-detection, we can see assets that aren't running as designed, in a very scalable way. What it really comes down to is the reason you get the hot/cold calls is because assets aren't working as designed. If you can correct those, prior to the person realizing they're hot or cold, that's a touchdown for us. We really are making the invisible visible.
>
> Fundamentally what we're doing is replacing the retro-commissioning [we were doing] and that's how we sold [the investment to management]. We're taking something that we did manually for a number of years...the challenge for us was we could only touch 200 [physical] assets a year, that's all we really had time for. Now that software's an enabler, we can touch 45,000 assets [the entire Redmond campus] in a year, and go deeper and wider [with our analyses and responses].[29]

Smith says, finally, "Give me a little data and I'll tell you a little. Give me a lot of data and I'll save the world." He believes that his approach offers guidance to anyone responsible for managing a large portfolio of buildings or facilities.

There are many smart building technologies that can be incorporated into a new green building rating system, so a daily score can be reported to management—with results for all buildings measured against a company's sustainability goals. In this way, a dynamic score can provide a valuable management tool now sorely lacking in static assessments provided by current rating systems.[30]

Few building portfolios are at Microsoft's scale, but *every* building can be managed better with off-the-shelf software that costs a pittance (less than $5 per month per metering point or key variable) and with Lucid's BuildingOS, all for less than $1,000 per year in total.[31] You can measure energy use, water consumption, waste generation and recycling, and you're going to spend less than $1,000/year to put this information on a software platform where it can be analyzed, visualized, compared with norms, reported to management, etc.

You've probably realized by now that I think a cloud software-based approach to smart building should form the core of future systems for certifying existing buildings. Every cloud-based building platform now on the market includes an ability to put energy data into an Energy Star score, and every platform includes an ability to upload data from many disparate sources: electric utility data feeds, monthly invoices from waste disposal and other vendors, quarterly employee commuting behavior surveys, etc., and display it on a Big Data platform, along with the ability to "normalize" data for occupant load, weather, building use, etc.

If you can measure, analyze and visualize ("see") data, software can spot anomalies immediately, track progress toward defined goals, immediately send SMS messages to everyone responsible for a given issue, and generate reports for facilities, upper management and outside stakeholders. Visual management tools such as heat maps (Figure C.6) allow a building operator to review a month's worth of electricity use data in 15-minute intervals. Anomalies in energy use can be immediately spotted by color contrasts and then further tracked by day and time. Such tools will be in the toolkit of all building and portfolio managers in the near future. Their availability represents an amazing opportunity to create new green building rating systems for existing buildings that have considerable value as management tools.

Rekindling the Green Building Revolution

The green building revolution needs to be rekindled. Current systems all lead to an evolutionary "dead end," a Neanderthal rating system with increasing complexity, high cost and limited adoption. Redesigning our rating and certification systems to take off from the intersection of the Internet of Things and Big Data analytics offers the potential for massive market penetration for green buildings and future certification systems over the next decade.

It's just plain stupid for green building rating systems not to use technology that we all carry in our pockets and purses, not to recognize how Moore's Law impacts our ever-increasing ability to manage data, and not to recognize that building owners have limited resources to assess building performance and to cut carbon emissions.

Why don't we redesign our systems so building owners will effectively own them and see value in using them? Why don't we admit that current rating systems effectively only serve the high-profile "1 percent" and instead create systems designed from the inception for mass adoption?

Leading-edge cloud-based software platforms point the way to *a data-drenched future*, one that offers potentially huge reductions in carbon emissions and building environmental impacts because everyone knows what's happening on a real-time basis and gets actionable information.

In our concluding chapters, we'll discuss how to leverage cloud technology, Big Data, the Internet of Things, social media and government action to address carbon emission reduction through green building rating and certification programs.

Then we'll offer a scenario-based approach for making appropriate decisions to create a far more desirable future. Finally, in the Epilogue, we'll sketch some promising aspects for the future of green buildings, with a look at new technologies emerging that will make buildings in 2020 look dramatically different from even 2010.

PART IV

THE FUTURE
OF GREEN BUILDING

CHAPTER 15

Reinventing
Green Building

What we want to do is make a leapfrog product that is
way smarter than any mobile device has ever been,
and super-easy to use. This is what the iPhone is. OK?
So, we're going to reinvent the phone.

Steve Jobs, 2007[1]

In pondering ways to replace the jury-rigged LEED system with new approaches that would have broader appeal, I looked for analogies far afield, including in the world of fine art. One that especially appealed to me: use sculpture as a model instead of painting.

An Analogy—Painting vs. Sculpture

Consider LEED like painting: start with a blank canvas, layer on green paint, one coat after another, somewhat like a Jackson Pollock painting (but without its redeeming fractal characteristics), pouring gallons onto a blank canvas, throwing red and brown on for effect, until the original straightforward idea—green buildings embodying best practices—becomes lost or very muddled. In our view, Reinventing Green Building should take a sculptor's approach (Figure 15.1): start with a marble block, i.e., one containing all good ideas for green building and then chip away at anything that doesn't directly deal with the "Big Three"—energy/carbon, water and waste—until what's left is a clear image, such as one we've all admired for 500 years, Michelangelo's *David*, a work of art almost everyone appreciates and understands.

Designing a Disruptive Innovation

If we're going to replace LEED as the major green building system in the United States or, alternatively, offer a green building system that has the potential for large-scale adoption by "the other 99 percent," what should the replacement look like? LEED is ripe for disruption by what Harvard Business School's Robert Solomon calls a Minimum Viable Product, containing "the minimal set of functionality that a user would find useful—and is willing to pay for."[2]

FIGURE 15.1. Future Green Building Rating Systems Should Be More Like Sculpture. Credit: © iStock.com/tunart

A true MVP or disruptive innovation may look like a "leapfrog product," way smarter than LEED and super-easy to use. It might be an iPhone compared to 2006's flip phone. Remember the Motorola *Razr*?[3] It ruled the mobile phone world in 2006 and yet became obsolete by 2008 after the iPhone's 2007 introduction. Or, it might meet Clayton Christensen's definition: a disruptive innovation is one that offers "simple solutions with less functionality and much lower prices than established competitors."[4] There are many situations where new technologies have overtaken and completely replaced legacy systems in short order. Think of examples like Uber and Airbnb. It's clear that incremental improvements won't do the trick, and that we can (and we must) reinvent green building with a far more disruptive approach!

In the last chapter, we argued for three key characteristics in a new system designed to meet most buildings' needs, something that bears repeating here. To meet market needs, a successful new system must be:

1. **Smart** (readily incorporating new technologies for building design and operations).
2. **Simple** (easily implemented by building operations personnel).
3. **Sustainable** (it must deal with key sustainability issues).

But in addition, the system must address four other key issues, by being:

1. **Owner-friendly:** our future green building system must clearly meet building owners' needs for sustainable operations, not the ambitions of some in the environmental community to create the ultimate green building "leadership" system.
2. **Cost-effective:** the future green building system must meet cost-effectiveness criteria for building owners and facility managers, in that it must deliver better results than they could achieve on their own.
3. **Built for rapid uptake:** the future green building system must be able to certify 25 percent of US building area during the next ten years, which means it must be embraced immediately by end users and not have to be "sold" to them by architects or consultants.
4. **User-friendly in its delivery:** the future green building system must deliver results quickly and fairly and should not depend on extensive third-party reviews, 700-page technical manuals or thousands of arcane interpretations from a remote bureaucracy.

Let's look at the three primary characteristics of a new system in detail, starting with smart, which includes our current understanding that all buildings are or can be remotely managed using low-cost, cloud-based technologies.

Smart

We know that cloud-based data collection, analysis, visualization, fault detection and diagnostics, and comparisons among buildings in a portfolio are now readily available for all buildings with IP-addressable building meters and automation systems. Such systems can easily be fitted to existing buildings with analog-based building management systems.

Smart building technologies already exist that can incorporate a new green building rating system directly onto a data platform, so a score can be frequently reported to management and outside parties, with results measured against an organization's sustainability goals and third-party criteria.

Simple

I have argued elsewhere[5] that ultimately we need green building rating systems that focus on absolute performance, not relative improvement or relative efficiency, in order to meet carbon emissions reduction goals. After all, nature doesn't care much about our relative improvement in reducing CO_2 emissions, it just responds to absolute amounts of atmospheric CO_2, and over the past decade these levels rose annually by 2 ppmv (parts per million by volume).[6] Within 10 to 30 years, depending on trends in emissions, we could reach a tipping point at which global climate change becomes in essence irreversible.

The problem with a relative-score approach is obviously that it doesn't evaluate buildings against the ideal: zero net energy use. It would be more valuable to set a goal for energy use for each building type (and size) so zero net energy would be included as a possibility, and then to use a rating scale such as that originally proposed in 2009 by Charles Eley with his Zero Energy Performance Index, shown in Figure 15.2.[7] Eley explains the origin of the concept in this way:

> The concept of zEPI sort of sprung from frustration. Everyone was making claims: "I'm x percent better than code." My first question was always "Which code?" and the second question was "Are you counting all the energy in the building, or just the energy that's regulated by the standard?" So

those two questions triggered my thinking about zEPI. The idea was to come up with a stable metric that would not change over time. There are two ends of the zEPI scale. One end is a zero net energy building, which is a pure concept. For the other end of the scale we decided to set it at the same level as the median building represented in the 2003 national CBECS database, which aligns with the median energy performance of a typical building at the turn of the millennium.[8]

Any green building rating system that assesses the energy performance of buildings should want to use a stable metric that doesn't change over time. Eley believes that this concept has technical validity for all buildings and also great intuitive appeal:

The beauty of zEPI is that the concept is exactly the same for new buildings or existing buildings. It's essentially the ratio of a building's energy use to the energy use of a similar turn-of-the-millennium building in a similar climate, with a similar use. So for a new building you would determine

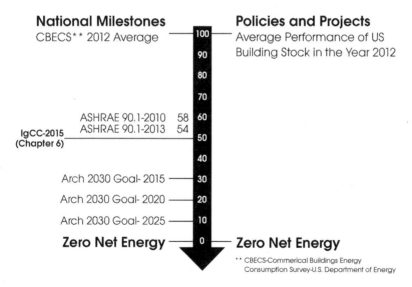

FIGURE 15.2. Zero Energy Performance Index (zEPI).[9] Credit: © 2015 New Buildings Institute

energy use on the basis of modeling because there's no history, and for an existing building you would calculate the ratio based on utility bills or other empirical data.

If future green building rating systems start with *zero net energy* as the ultimate goal, here is a tool that's readily available. It allows you to evaluate energy use in all buildings on a consistent basis and then award energy points based on that result.

Sustainable

People will likely argue forever about which sustainability concerns should be incorporated in a green building rating system, but marketplace evidence from the recent past argues decisively for keeping them to a small number. What stands out for me are these key performance indicators, or sustainability KPIs:

1. **Energy Use:** With a "zero net energy" goal (setting aside the likelihood that very few buildings will or can ever become "plus-energy"[10] buildings), this will include both direct combustion (natural gas or diesel for water heating, for example) and indirect combustion (electricity), while incentivizing onsite production from renewables or biomass boilers.

2. **Water Use:** Recognizing that we are entering a time of global water scarcities, brought on by population growth, climate change, increasing water footprints from agriculture, cities and industry, we need to reduce water use to an average achieved by lowest-using developed countries.[11] (I do *not* support aiming for zero net water use in buildings, for reasons given in Chapter 12.)

3. **Waste Diversion:** Most US urban waste recycling systems seem to have peaked at around 35 percent waste diversion from landfill.[12] It seems reasonable then to assess green buildings by starting with 50 percent diversion as a higher goal and embracing a zero waste ideal for waste generation and disposal.[13]

4. **Scope 3 Carbon Emissions:** Scope 3 emissions are essentially "induced" emissions from corporate travel, freight deliveries and employee commuting. All can be easily tracked on a monthly basis from vendor invoices (which can be formatted for upload to FTP

sites and then "grabbed" by dashboard APIs) and quarterly or semi-annually from employee surveys. Our goal is to encourage companies to reduce Scope 3 carbon emissions to zero through many means, including purchasing carbon offsets.

5. **Ecological Purchasing:** While it may be limited initially to office products and similar items bought from a handful of vendors, this measure would provide useful data. Some larger US office supply companies such as Staples and Office Depot have clear and valuable programs for labeling ecological products, which can then provide input to monthly invoices for determining the total percentage of purchases that meets these criteria. The goal is clearly to get 100 percent ecological purchasing for ongoing operations.

A rating system focusing on these KPIs might look like that shown in Figure 15.3, based on criteria shown in Table 15.1.

TABLE 15.1. Sustainability KPIs—A New Rating System

	Points:	Min ◄————————————————► Max	
Energy	50	30% below 2012 CBECS	Zero net energy
Water	20	30% below US average	90% below
Waste	10	50% diversion	100% diversion
Scope 3 Emissions	10	50% below US average	Zero net carbon
Purchasing	10	50% of purchases/year	100% of total

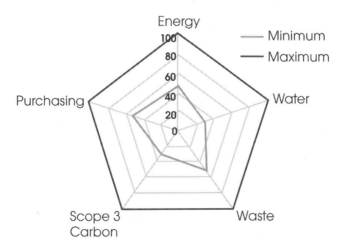

FIGURE 15.3. Schematic Rating System for Sustainability KPIs

This approach is straightforward, works for just about every building type, is easy to explain to just about anyone and creates a compelling reason to get everyone on board for achieving sustainability goals!

Design on the EDGE

The International Finance Corporation (IFC) of the World Bank group created the EDGE (Excellence in Design for Greater Efficiencies) software to help (eventually) 100 emerging market countries design "20-20-20" buildings, which save 20 percent of energy and water use compared to a base case, and which reduce embodied energy in building materials by 20 percent. The software works for five common typologies: homes, offices, retail, hospitals and hotels. It is free to use and certification is available for a small fee, using GBCI or local auditors from IFC partners in each country, except for India, which will certify solely via the GBCI.[14]

These projects are well on the road to being smart, simple and sustainable, using easily accessible design tools. The EDGE certification program is currently available in Costa Rica, India, Indonesia, South Africa and Vietnam. This is another good idea that addresses a global issue: 70 percent of the world's population will be living in cities by 2050, so we should be trying to make new building designs as resource-conserving as possible.[15] However, EDGE has yet to demonstrate that its certification costs are low enough to encourage widespread use in the emerging market economies that it targets and that the buildings it certifies will achieve the 20-20-20 goals in practice.

Focus on Continuous Improvement

We could make it even easier by focusing on continuous improvement: my friend and colleague Steven A. Straus, CEO at Glumac, a building engineering firm, argues for something anyone can understand: *1-2-3! Sustainable!*

In his view, we need to encourage people to commit to regular but small annual improvements, so in ten years these small gains would add up to large numbers. This approach would firmly embrace and support a continuous improvement culture in many companies, government agencies and other organizations. We could call this program simply:

One-Two-Three-Sustain-abili-tee!

Program users would commit to *annual measurable and reportable improvements* in these three variables, measured against the *current* baseline (or perhaps an average from the most recent three years):

- **Waste Generation** (Diversion from Landfill): **1** percent increase each year in recycling (this leaves open an opportunity to reduce *absolute* levels of waste sent to landfill).
- **Water Use: 2** percent reduction in water use, which can be accomplished through several means: plumbing fixture replacement, change to native or adapted landscaping, increase in concentration cycles for a cooling tower, etc.
- **Energy Use: 3** percent reduction in annual site energy use, which can be accomplished through such means as installing solar power, switching to LED lighting fixtures, introducing smart building technologies such as *predictive optimization* (automated setpoint reduction), etc. Because we are referencing nominal reduction from an established "year 0" baseline, annual reductions would be constant and the organization's ongoing challenge would be to keep finding opportunities for further reductions.

Using this approach, after ten years, an organization would have reduced waste generation by 10 percent from current levels, water use by 20 percent and energy use by 30 percent. Isn't this a better approach than insisting that as a prerequisite for an existing building to be called a green building, it already has to have a high Energy Star score or (in LEEDv4[16]) demonstrate a 25 percent energy use reduction against its own prior three-year average?

Given that President Obama's 2015 executive order[17] mandates that federal agencies reduce energy use by 2.5 percent per year and water use by 2 percent per year, it's clear that these goals are attainable for most buildings and only await some organization willing to put

Straus's "1-2-3 for Sustainability" into a rating system with a marketable label.

Reaching Beyond the "1 Percent"

David Pogue of CBRE is a well-respected commentator in the building management industry. He points out several challenges in reaching beyond top-tier buildings:

> In my view *the first movers have moved.* The first movers are the ones who have motive and money, and they have staff and stakeholders supporting, even demanding they move. So they have. You could almost say that that's a bit of a wholesale market, because they often are certifying whole portfolios and they're engaged.
>
> The rest of the market, which represents the bulk of the buildings and square footage, is more of a retail market. It's going to change one-by-one, because they don't have multiple buildings. They also don't have many of those other motivators, so you've got to appeal to Mr. and Mrs. Building-Owner in every small town everywhere, and figure out how to message that and make it matter to them.
>
> I'm perplexed on how you make that happen. How do we move from a wholesale to a retail marketplace? Many of the firms engaged in energy efficiency or certification are making money on a wholesale play. They're dealing with the big properties and the large firms, and I don't know how you can get the money or the resources or the marketing to do the 25,000-square foot buildings, because there's not any money there.
>
> The choice currently is to make one big deal with one big owner, or a thousand little ones. Someone needs to find a way to aggregate a number of the smaller ones. There will be a time I guess, when we can simplify how you do that, but right now I think that's the disconnect [with reaching the larger marketplace].[18]

To reach owners of smaller buildings, Pogue thinks there may be a role for government and for utility incentives, especially for energy efficiency upgrades, as well as government mandates, e.g., for performance disclosure to the public, especially to tenants and future buyers.

Conclusion: A New Green Building Manifesto

If you've read this far, you probably agree that we need to reinvent green building so it can reach its full potential and not get stuck with the "1 percent," where it currently stands. We need to approach the problem with new ideas and new intentions. When innovators started designing green building rating systems in the 1990s, they had many contrasting and not wholly reconciled environmental goals in mind.

Developers of LEED (and other systems) *never* went to the larger user base to ask *what they wanted* in a green building standard. I'm not sure that would have been very useful then, because those (few) users who cared about sustainability wanted to be told what to do by experts.

Times have changed. Building owners are now much more sophisticated about green building and sustainability—go to any conference and listen to building owners, and you'll agree. But these owners need a standard that will meet their business needs and be implementable by property managers and facility managers, while still delivering reductions in energy and water use and reducing overall carbon emissions.

Our new green building manifesto needs to take into account this new attitude and orientation, along with building owners' and facility managers' more sophisticated understanding. We need to rethink LEED's "everything but the kitchen sink," splashing-more-paint-on-canvas approach and refocus our energy and attention on climate change, water scarcity and waste generation.

That is our task ahead. We may want to keep the complex LEED rating system for use in large new buildings designed for and operated by high-end building owners who value a more comprehensive and challenging certification regimen. So in that sense, LEED, BREEAM,

Green Globes, Green Star and similar systems have a valuable *but limited* place in the green building certification market.

But the "other 99 percent" need a system that makes sense to them, one they can implement (as we showed in Chapter 12 with JLL's Green + Productive Workplace system) with minimal staff time, and get not only an evaluation of where they stand, but also clear direction as to how to proceed in the future. In our concluding chapter, we present five scenarios for evolving new directions. Take a look!

CHAPTER 16

Green Building Futures

Two roads diverged in a wood, and I—
I took the one less traveled by,
And that has made all the difference.

Robert Frost[1]

Green building certification in the United States is at a crossroads; LEED's annual use has flat-lined overall since 2012, with projects primarily limited to top-tier commercial real estate, corporate buildings, selected government bodies and high-end universities. These are important but limited market segments. To cut carbon emissions from buildings on a massive scale, LEED simply can't get the job done as it's now constituted, and so far no competing certification system has shown an ability to reach significantly beyond these limited market segments.

I wrote this book to help us decide which direction the green building movement should take. I've shown you "the good, the bad and the ugly" about our current direction and have tried to profile honestly the compelling need for major changes in our approach to cutting carbon emissions from buildings. At the beginning of the book, I issued a "call to action" to Reinvent Green Building. If you've read this far, you will hopefully agree with me that the time is long overdue for a critical appraisal of our success in mainstreaming green building, especially in getting the broad market of building owners to adopt and use certification systems. In this chapter, we look at some of the choices that lie ahead.

Choices need to be made soon and decisions taken about the future. If we refuse to look at our situation dispassionately and to make painful but necessary changes, we're choosing to remain with the status quo—even if it's not working. That's the worst choice because we can't keep improvising solutions that will cut carbon emissions from buildings faster and far more dramatically than we have to date. And we can't give up the effort, either. If one tool becomes too dull to make the proper cut, the choice is either to sharpen it or pick up another one better suited for our task.

As the late Yogi Berra, America's preeminent practical philosopher, once said, "When you come to a fork in the road, take it."[2] It's time to take a new road, to make a new path that leads demonstrably and decisively toward a low-carbon future for ALL buildings.

If, as we have demonstrated, LEED has peaked in popularity, what are available options to create a winning customer experience strong enough to reinvigorate the green building movement? Let's set an ambitious goal: green building over the next five years will resume the rapid growth of the 2006–2010 period, and by the early 2020s will be well on its way to delivering on a new promise: achieving the universal application of key green building measures in design, construction and operations.

Five Scenarios

Scenario building is a well-established approach to dealing with the future.[3] Considering new ways to develop and analyze alternative near-term futures, I will put forward five basic scenarios for green building's future (Figure 16.1). After reading this chapter, you may be moved to come up with your own scenarios for further analysis.

1. **Business as Usual.** In this scenario, nothing much changes. USGBC decides to go with LEEDv4 (perhaps with incremental changes) and ignore the naysayers. This is currently the most likely scenario. However, the analyses in this book and thousands of professionals' practical experience say that it won't generate measurable gains, let alone massive transformation. In this sce-

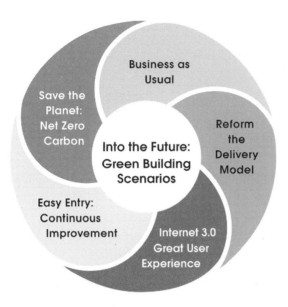

FIGURE 16.1. Five Green Building Scenarios

nario, other rating systems continue to compete with LEED for market share in the nonresidential market. The residential market continues to show a preference for multiple green home rating systems, with LEED making up 25 percent, the somewhat weaker NGBS 25 percent, and the balance divided among regional systems at varying levels of rigor and results. In this scenario, the LEED label in the residential market would continue to represent only one "best in class" view, primarily for the multifamily market.

2. **Reform: An Internal Revolution (Revolt) at USGBC to Adopt the BREEAM Model.** In this scenario, USGBC decides to stick with LEEDv4 or some more realistic, slimmed-down rating system, but dramatically reforms both the content and delivery model. It decides to quit relying on the "more than 300 full-time GBCI technical experts and consultants"[4] and move toward the BREEAM model of independent assessors who can deliver a valid independent certification at a total cost 50 to 90 percent below today's

costs. In this scenario, USGBC also creates a separate system to address cost/complexity issues and creates a new brand that addresses especially those market segments that have resisted going "all in" for LEED.

3. **Internet 3.0: An Entirely New Approach that Focuses on Great User Experience.** In this scenario, the focus is on the user experience, and the system lets users contribute to a significant degree to developing credits and certification protocols. To get this done, a major organization, one without USGBC's institutional inertia, would need to deliver a technologically-enabled product development and delivery model, one that works on a smartphone and sits on a cloud-based platform. (Google/Alphabet may be a logical candidate for this assignment.) This scenario incorporates Key Performance Indicators for sustainability outlined in Chapter 15, and lets users assess energy use, water use, waste recycling, purchasing and Scope 3 carbon emissions on a platform that easily enables third-party app developers to create ratings for each building industry segment.

4. **Save the Planet: Zero Net Carbon.** In our fourth scenario, we abandon all efforts to load up green building's definition with anything other than reducing buildings' carbon impact; we reconstruct all rating systems to reflect the carbon life-cycle, in direct energy use and indirect use in transportation, building services and building products, using Life-Cycle Assessment (LCA) analyses for building materials, water use, etc. The system would consider Scope 1, 2 and 3 carbon emissions, along with "Scope 4" carbon emissions that include life-cycle carbon impacts stemming from all sources related to building construction and operations.

5. **Easy Entry: Reward Continuous Improvement.** There should be a place to recognize committed and prolonged efforts (especially by government agencies and large institutions) to deliver annual improvements that cumulatively, over a 10-year period, create meaningful total reductions in carbon emissions, without requiring participants to meet minimum performance standards. Like

the mythical Hotel California, you could check in easily, but once you made a public commitment there would be no way to check out, to return to your old way of doing business.

Let's examine each scenario in turn.

Scenario 1: Business as Usual

This scenario is well underway, as the LEED label is a well-established brand, but the downside is that most projects either just don't bother with certification or pick and choose from a menu of green design measures or operating activities, then call it "certifiable" or "LEED-Platinum equivalent." Within this scenario, USGBC continues to promote LEED as a global brand but faces increased competition from local brands that may be cheaper, easier and faster in use, including GRIHA[5] and India Green Building Council in India, DGNB in Germany, HK-BEAM in Hong Kong, Green Mark in Singapore,[6] a resurgent BREEAM in Europe and Mexico, alternative rating systems in Brazil, Pearl and QSAS in the Persian Gulf countries and so on. LEED would retain its cachet, but its market impact would remain limited mostly to high-end, high-profile projects.

Given that USGBC basically hasn't changed how LEED works for 15 years (yes, there are newer versions and more rating systems, but the system's core structure and approach from 2000 remains in place) this is the most likely scenario. Using this approach, LEED will continue to lose US market share to the "none-of-the-above" alternative and the situation is ripe for a strong competitor to enter the market with a better product, user interface and user experience.

In the residential sector, the battle between rating systems is already over and the verdict is that they will mostly stay small. In the single-family residential market, regional brands will continue to grow—Earth Advantage in the Pacific Northwest, GreenPoint Rated in California (more than 38,000 homes certified),[7] EarthCraft Homes in the Southeast (more than 30,000 homes certified),[8] Austin Energy in Texas (about 12,000 homes certified),[9] etc.—and the National Green Building Standard will remain a significant national competitor. The only area where LEED has good opportunity to gain traction is the LEED Multifamily product, especially since most US apartments built in 2015 are high-end and therefore may find some marketing advantage in appealing to "green" consciousness held by potential buyers or renters with higher incomes.[10]

This most likely scenario has one overriding problem: *It meets a standard definition of being crazy, i.e., continuing to do the same thing over and over and expecting different results.* It is likely to lead to USGBC's further decline, as US certification revenues continue to decrease, USGBC continues to post annual operating losses, and as LEED's global audience starts to rebel against high costs and slow-motion bureaucracy.

Scenario 2: Reform—LEED Adopts a New Format and Delivery Model

In this scenario, USGBC and GBCI throw in the towel and admit that LEED's current approach is far too complex, opaque and costly for significant market uptake. Instead, it applies Occam's razor[11] to the existing LEED-EBOM system and reduces the LEED-EBOM system to five credits: energy use, water use, waste recycling, sustainable purchasing and Scope 3 carbon emissions, as described below.

Replacing the current legions of staff and reviewers, LEED changes the delivery model to allow LEED APs with specialty (similar to Green Globes Assessors and BREEAM Assessors) to have FINAL AUTHORITY over project certification, with a mandate to render a judgment on project certification within four weeks (or less) after receiving all project information. Documentation could be done easily using a BREEAM (or Green Globes) style of online questionnaire, and the assessor would be required to use Skype (or a similar approach) to hold a certification review meeting with the project team, so there would be no need for a site visit. GBCI would audit, at most, every 10th project in the beginning and every 25th or 50th project after a while, so there would be no need for a large headquarters staff or legions of reviewers.

GBCI would also change the LEED-EBOM format. In this scenario, all prerequisites would be abandoned as antiquated, unnecessary and costly, and the point system would be adjusted to focus attention on just five sustainability KPIs: energy, water, waste, materials purchasing and Scope 3 carbon emissions.

In particular, following the suggestions in Chapter 15, the new format would include:

1. 100-point scale (no more "innovation" or "regional priority" credits)
2. Adoption of Architecture 2030 as the basis for the sole energy credit, given 50 of the 100 points and weighted more heavily based on how close an existing building comes to meeting or exceeding Architecture 2015 goals. In 2020, this system's energy goals would automatically ratchet up to Architecture 2030's goals for 2020. On-site renewable energy production would count toward meeting the 2030 goals.
3. Water efficiency could be put on a similar scale to Architecture 2030's energy goals, by working with the Alliance for Water Efficiency to create the measurement scale and aiming at "best in developed world" water use. This credit would be given 20 points.
4. Waste reduction toward zero waste generation, again weighted more heavily above 75 percent and given 10 points.
5. Scope 3 carbon emission reductions for employee commuting, starting at 25 percent and weighted more heavily above 50 percent. Credit would be 10 points.
6. Sustainable purchasing using apps readily available from major office supply vendors. Goal would be 100 percent, credit starts at 75 percent. Credit would be 10 points.

That's it—five credits, 100 points: smart, simple and sustainable. Get rid of everything else that's driving LEED-EBOM use into the ground, including having to use expert consultants and assembling documentation for such "best practices" as integrated pest management and green cleaning which are irrelevant to cutting carbon emissions. Then make a deal with one or more major software platforms such as Lucid or Switch Automation to host the information, open up their API to LEED-EBOM software developers, and give building and portfolio owners access to a tool that provides a continuous reading of the LEED-EBOM score, as well as the other beneficial features of these dashboards.

Of course, they'd still have to pay for use of the platform, but think about having many more sales people offering LEED-EBOM to all the building owners they're pitching to use the platform, maybe even including it as a standard feature in the annual fee. There could even be a "freemium model" that allowed tracking the LEED-EBOM score at no cost, but did not include certification.[12]

One more thing: it would only cost $2,500 for each of the first 10 buildings, $2,000 for the next 40 and $1,500 for everything over 50. If a retailer with 100 stores wanted to track sustainability in all stores, it would cost only $180,000—$1,800 each. The program requires re-certification every three years, so the initial three-year certification cost would only be $60,000 per year, $5,000 per month and a little staff time to set up the system. *As a building owner, if I could get a 100-building portfolio accredited by LEED for $50 per month per building, why* wouldn't *I want to take part?* This system could be immensely attractive for convenience stores, restaurant chains, hotels, commercial offices, universities, K12 schools, etc.

What about the economics? One hundred new 100-building portfolios per year would yield $18 million in revenues to GBCI and would bring 10,000 buildings into the LEED-EBOM system each year instead of the current 500 to 600.

This new existing-building system would be marketed as a companion to the rest of LEEDv4 and would be specifically aimed at the more cost-sensitive segments of commercial and institutional building markets, i.e., almost everyone but high-end office developers. What's wrong with having more than one product in the marketplace?

Why do this? It's vital for USGBC and GBCI to expand the market and to reignite growth that has stalled for several years as more building owners adopted the "thanks, but no thanks, I'll do it myself" approach. GBCI isn't going to grow by becoming the certification arm for other systems such as GRESB and WELL, which both target the same market where LEED is already well established: commercial offices.

GBCI should adopt this approach *before* implementing LEEDv4 EBOM as the only available system in October 2016, if only for self-preservation. Certification revenues are likely to decline further, particularly if a strong competitor (such as BREEAM) enters the US market and targets high-end portfolios that need to report their results to a global audience via GRESB or similar systems.

There's only one thing holding back this reform: *Stinkin' thinkin'*[13] that LEED-EBOM has to look and operate similarly to the LEED system for new construction. This approach is so simple, direct and easy to implement that I'm almost ashamed to present it in this book, since GBCI could easily have come up with this idea on its own—but it didn't!

Scenario 3: KPIs and Internet 3.0

The technology-enabled option for green building certification, presented throughout this book, when combined with a sustainability KPI approach, offers the potential for green building to ride the wave of technological change upending the real estate and building operations industry, in the same way that the iPod upended the recording industry, Uber and Lyft the taxicab industry, Airbnb the hotel industry, the iPhone the mobile phone business, and the iPad the desktop/laptop computer industry.

There's too much money, easy-to-use technology and software smarts pursuing building owners, developers, government staff, operators and facility managers to cling to a green building certification model developed during the 1990s "desktop" era.

Implementing this scenario will more than likely require a new organization, one aiming squarely at the "early majority" segment and one committed to "crossing the chasm"[14] from innovators and early adopters to broader market acceptance.

The chasm-concept's founder, Geoffrey Moore, describes the issue as finding buyers who "are 'pragmatists in pain,' stuck with a problem business process and willing to take a chance on something new, provided it is *directly focused* on solving their specific use case."[15] That surely describes the conundrum for today's green building audience. They want to do the right (sustainable) thing, but they're not willing to pay LEED's price to have GBCI certify that they did it.

This new organization should have the financial strength needed to launch a new green building rating system and enough savvy to craft a brand name that will resonate with the user market. It also will need to design a way to stay alive financially until its brand can take hold. If we take G+P as a model, then at a $1,500 certification fee this program would require about 7,000 annual certifications to build a $10 million business that could afford a staff large enough to sell and administer such a program. It's clear from LEED's initial success that there is certainly such a large user base potentially available.

Scenario 4: Zero Net Carbon

In this scenario, an organization such as Architecture 2030 or New Buildings Institute (to name just two—there are many more) takes the zero net carbon opportunity seriously enough to launch its own certification product, one that will drive low-energy buildings into the mainstream. One could even envision an adventurous and technologically sophisticated company such as Google/Nest or Tesla/Solar City taking on the challenge, both to add to their sustainability investments and to bundle hardware and software technologies into marketable and deliverable packages for many buildings and homes. Support could also come from the Carbon Disclosure Project, the Carbon War Room, or another well-financed organization aiming squarely at the climate change issue.

The Carbon War Room has significant credibility and a clear vision:

> Our vision is a world where over $1 trillion invested in climate change solutions is an annual occurrence, not a historic milestone. In this world, market barriers will not exist in any

sector where profitable carbon reduction solutions exist; and entrepreneurs who are passionate about preserving our planet's resources are simultaneously tapping into the economic opportunity of our generation.[16]

The problem with our zero net carbon scenario is that it needs a prime mover with financial strength, credibility and strong technology offerings to make it happen. We envision a major US nonprofit with strong carbon-reduction credentials and an entrepreneurial management team taking advantage of LEED's weaknesses in critical market segments such as smaller office buildings along with education, healthcare and retail building portfolios, to build a brand and a business that would eventually deliver carbon reductions along with green building certifications that many people want but are not willing to pay for at today's prices.

As an initial market entry point, 685 US colleges and universities that have signed onto the ACUPCC[17] (see Chapter 5) could form a solid market base for a zero net carbon commitment, one that would have widespread support among academics and students, alumni and local businesses alike. A successful university initiative could open the way for targeted K12, government, healthcare and retail initiatives.

Scenario 5: Reward Continuous Improvement

Founded on continuous improvement, a well-established philosophy and business practice, this scenario offers an entry point to sustainability, using an organization's serious multiyear commitment to improving current operations by a fixed amount each year. We envision large organizations such as the US military with established command-and-control structures, each with hundreds (if not thousands) of buildings and facilities (consider government buildings and also the retail sector) becoming the first to want to make this sustainability commitment a reality.

In this scenario, we envision grafting the continuous improvement idea onto existing programs for such systems as the International Council of Shopping Centers' (ICSC) Property Efficiency

Scorecard,[18] perhaps in cooperation with a sophisticated energy or sustainability-oriented nonprofit.

Within the US government, President Obama's 2015 executive order, mandating sustainability improvements for federal agencies, ties in well to this scenario but needs an implementation program, perhaps through the well-known Federal Energy Management Program or the Interagency Sustainability Working Group,[19] aided by some technologically savvy help from the General Services Administration, the government's main property owner and operator.

Scenario Review

You can see that these scenarios are not mutually exclusive, but they do involve some clear choices about where we should place our energies and our business bets.

- **Business as Usual** is the clearest option for LEED, but it doesn't do much for everyone else. LEED would continue to promote itself as the "leadership" standard, applauded (by some) for its rigor and derided (by many others) for its rigidity, but would likely face a declining US market and growing international competition.
- **Reform** would require some genuine soul-searching from an organization that hasn't been noted for introspection or tolerance of dissent, but it offers LEED a one-time golden opportunity to reinvigorate its core audience and to reclaim a leadership role it may soon have to cede to other more nimble and customer-oriented organizations.
- **Internet 3.0** is already well underway in building operations, but most technology players aren't yet awake to the opportunity to graft green building certification onto their powerful platforms. It is not yet a reality in the residential market, which still lacks the resources of a major company to implement it. To craft a certification system to serve this vision would require an organization, new or existing, with enough financial strength to build a staff and create a brand strong enough to attract user interest and kick-start such a program.

- **Zero Net Carbon** has the greatest potential for immediate impact, not because it would certify a lot of buildings right away, but because it could be readily understood and supported by both building owners and the general public. It may be that the residential market is the best place to start, working with a consortium of leading national and regional homebuilders. What this scenario lacks right now is a major institutional sponsor with financial resources and a business mindset.
- **Reward Continuous Improvement** would produce immediate results over the next five years, and with the federal government pushing forward with a clear sustainability mandate from its CEO (President Obama), it could have a huge impact on US carbon emissions within ten years. That's worth doing, no matter what happens to green building rating systems.

Conclusion: Reinventing Green Building

This book is about choosing to "reinvent green building" by taking another path from the one that we've been on for the past 15 years. We have demonstrated, convincingly we hope, that our current path won't get us where we need to go. We hope that you will be inspired to question your own commitment to this approach to green building and to start demanding accountability for results from this movement's leaders.

We also hope that you will see that there is more than one scenario for a creating a sustainable future, one that marries green building's ethical component with sound business thinking, aided by adopting technologies we all engage with every day.

With this book, we hope to stimulate a long-overdue debate about better ways (and there are surely many more options than LEED) to secure a low-carbon future and a healthier, more environmentally responsive built environment.

Complex systems can't be changed easily. *It takes a system to change a system,* and that involves changes in thinking, market testing our assumptions, employing significant financial resources and maintaining flexibility to choose alternative paths if one approach isn't working.

We believe that new Internet technologies for monitoring and managing buildings offer a unique opportunity to reinvent how we approach green building, to design and deploy certification systems that will meet the carbon-reduction test. In the concluding Epilogue, a leading futurist, Timothy C. Mack, offers a glimpse into future green building technology, inspiring us with how quickly we can make changes, or more accurately, how quickly change is going to happen regardless of our readiness for it.

When it comes to choices for the future, I am an incurable optimist, believing with Winston Churchill (perhaps the 20th century's greatest optimist, leading England in facing down a seemingly omnipotent Nazi Germany when his country stood all alone in 1940 and 1941), that "A pessimist sees the difficulty in every opportunity; an optimist sees the opportunity in every difficulty."[20]

I also believe, with R. Buckminster Fuller, one of the leading 20th-century American designers and futurists, someone who embraced the future with great enthusiasm, "You never change things by fighting the existing reality. To change something, build a new model that makes the existing model obsolete."[21]

Only you and I can build this new model. It's our choice about how we do it. But it needs to be done NOW!

THE FUTURE OF GREEN BUILDING TECHNOLOGY

Green Building Technology to 2020

by Timothy C. Mack[1]

*We must test all intellectually respectable lines
of inquiry, while keeping in mind that, as the great
Danish physicist Niels Bohr said,
"It is very difficult to predict—especially the future."* [2]

Green building is not limited by any means to certification systems, the focus of this book. Clearly, green building is experiencing an *innovation explosion* that is transforming its future, including new building materials, new strategic building management tools and new construction technologies. But beyond some green building certification systems' inability to keep up with and adapt to these changing conditions and possibilities, the larger community of stakeholders lacks understanding about green technology possibilities. These innovations are driven in part by powerful technology convergences occurring within nanotech, biotech and materials sciences.

Innovation in green building is gaining speed, and will accelerate over the next half-decade and beyond, yielding significant implications for low-carbon building design in the green building industry. In this Epilogue, we'll focus on the future for green building materials, particularly those that affect air quality and reduce carbon emissions in their manufacture and use.

Changing Materials

- Air-cleaning paint, containing nano-scale titanium dioxide particles, is used to reduce harmful emissions from power plants and motor vehicles by interacting with light to break down nitrous oxide and volatile organic compounds into harmless substances. Adding silver nanoparticles to paint also prevents growth of mold, algae and bacteria.

- Green cement, developed in Germany several years ago, is a material that meets or exceeds ordinary Portland cement's functional performance capabilities by incorporating and optimizing recycled materials, thereby reducing consumption of natural raw materials, water and energy, resulting in a more sustainable construction material. Today green cement production accounts for 3.5 percent of global cement, but it has been forecasted to grow to over 13 percent of the market by 2020.[3]

- Fire-retardant insulation that can be produced from waste materials, such as shredded denim, plastic milk bottles, newspapers, agricultural straw, hemp and flax, to replace chemicals whose health impacts have been called into question.

- A Dutch bio-resin composite called Nabasco, which stands for **na**ture **bas**ed **co**mposite, is reinforced with natural fibers such as flax and hemp and used in panels produced with less energy and fewer chemicals than fiberglass but that are lighter and just as durable.[4]

- Smart electrochromic and thermochromic window glass can respond to environmental conditions to utilize natural sunshine and heat to offset the need for artificial lighting and artificial heating from HVAC. Suspended particle device (SPD) technology offers the highest switching speed plus maximum user control by using nano-scale particles and varying voltage levels.[5] The global smart glass market is expected to reach $700 million in annual sales by 2024.[6]

Accordingly, office buildings with these features can clean themselves, improve indoor air quality, and respond to sunlight by adjusting window tint. Such buildings aim to attract tenants willing to pay higher rents to get workplace productivity gains.

New Building Air Quality Technologies

- Materials originally designed to protect exterior surfaces from pollution damage (commercially known as TxActive) can also absorb pollution from the atmosphere and improve outside air quality through a photocatalytic process that utilizes UV in sunlight to break down and oxidize nitrogen and sulfur oxides as well as particulates and volatile organic compounds. Accordingly, the building coating improves ambient air quality in the surrounding neighborhood.[7]
- French researchers mimicked materials found in nature, such as the Spruce cone, that open and close according to humidity changes, thus enhancing mechanical sensing for controlling ventilation within office buildings.[8]

Reducing GHG Emissions

However, it is not just cutting-edge materials innovation that excites our imagination for what's possible in the coming decade. At present, 59 percent of Fortune 100 companies and nearly two-thirds of the Global 100 have set GHG emissions reduction commitments, renewable energy commitments, or both, including plans to have clean energy sources by 2020 meet 30 percent of demand.[9] Meeting this goal will require considerable near-term innovation in electricity generation and distribution.

Microgrids. As described in earlier chapters, one increasingly attractive goal is zero net onsite energy use. In this scenario, several connected buildings become an energy-self-sufficient *microgrid*, exchanging power among its members from diverse power sources with the ability to sell surplus energy to a local electric utility.[10] In Germany, where renewable energy already powers 30 percent of electric generating capacity, locally generated power upended the previous electricity supply system. Important questions of electric-grid reliability still need to be worked out, but it's clear that decentralized power generation has many practical, economic and strategic

FIGURE E.1. Microgrid with Solar PV at Camp Pendleton, CA. Credit: CleanSpark

advantages and will come into its own during the next decade. Figure E.1 shows an operational microgrid at Camp Pendleton Marine Base in San Diego County, California. Obviously, military installations are prime customers for future microgrids, as they need power 24/7, have multiple buildings with different time-of-use demands and typically have secure space for large PV installations.

Zero Net Energy Homes. At present, this challenge is less technical than financial, as zero net energy residences still cost more. KB Homes' *Double ZeroHouse 3.0* partners with SunPower for solar energy power and energy storage, with Ford for energy efficiency and Whirlpool for smart networked appliances, but it is designed for luxury markets.[11] As technology costs continue to fall, access to zero net energy housing is likely to expand beyond California.

One essential element to these homes is a 7 to 10 kilowatt-hour battery inside the house, serving both as a backup in case the grid goes down and leveling loads on the grid by charging up at night when electricity is cheap, and sending power into the grid during peak-use periods when power is more expensive to generate. While some US utilities allow a homeowner to run his or her meter *backward* at times, very few are willing to pay (or to keep paying) full market value for excess power beyond what a home uses over a year, so storing a home's solar-generated electricity onsite might soon become the more attractive option.

Nest Labs (owned by Google) solved a persistent challenge to acceptance of new technologies: ease of understanding and operation. The Nest Learning Thermostat is self-programming through its own experience heating and cooling a home. It claims to pay for itself in two years (even though it is not inexpensive); some US utility rebates for this technology are already in place.

Tesla is coming at energy reliability from the other end, with its Powerwall system, utilizing advanced battery technology to bring uninterruptible power supply (UPS) to homes.[12] Partnered with solar technology systems, lithium-ion cell systems provide one way to go (also with possible utility rebates), but hydrogen fuel-cell systems are also promoted by the US Department of Energy, as at present that cost is five times lower than lithium-ion batteries.[13] However, Tesla is definitely committed, as by 2017 it will be the world's largest producer of these batteries.[14]

Changing Green Building Materials Markets

Many still believe that costs associated with high-level green building construction are more than those for normal buildings, but the truth is that cost increments vary considerably, from nothing to five percent or more. Most studies show considerable future market growth for green products in commercial construction; typically costs come down to conventional levels as the installed base grows. Even while certification systems have hit the wall, the market for green products and services continues to grow, owing to stringent regulations, policies and incentives adopted by North American and European governments, plus some in Asia and the Middle East, for shifting toward green building technology. Future market growth for green building technologies and systems, including using green materials, will be driven by policies and regulations that prioritize energy efficiency and sustainable outcomes, develop new certification approaches for green buildings, cost reductions for green materials, and growing consumer demand for healthier buildings and healthy building materials.

According to Navigant Research, the global green building materials market will grow from $116 billion in 2013 to more than $254 billion in 2020. These green building materials will range from traditional materials revalued upward for their minimal environmental impacts to new technologies that enable better passive and active building performance. According to Navigant's study, "Materials in Green Buildings,"[15] future market growth for green buildings and green materials use will be driven by policies and regulations that prioritize energy efficiency and green design, expanding voluntary certification programs for green buildings, cost reductions for green materials, consumer demand and growing evidence that green buildings confer quantifiable market advantages. Europe may constitute green building's largest regional market for new materials.

Green Cement: Low-Carbon Outcomes from Traditional Systems

At present, worldwide cement production represents around four to five percent of human CO_2 emissions;[16] of this total, 60 percent is used in buildings and related construction.[17] In Switzerland, researchers at Ecole Polytechnique Federale de Lausanne (EPFL) received funding to focus on developing and testing new blends of low-carbon cement. This new green cement has the potential to reduce concrete's carbon footprint from construction activity by 40 percent.[18] Cement use continues to grow in rapidly developing countries such as India and China, with the global market expected to grow 66 percent between 2011 and 2016.[19] Developing countries now consume more than 90 percent of global cement production.[20]

EPFL's green cement uses calcined clay and ground limestone in larger proportions than conventional cement. Aluminates from clay interact with calcium carbonates from limestone and create a cement paste that is less porous and stronger than traditional cement. Portland cement is cheaper to make but is more caustic, and contains toxic ingredients like silica and chromium.

CarbonCure™ provides another approach: CarbonCure retrofits concrete plants with a technology that recycles waste CO_2 into affordable, greener concrete products.[21] With this approach, instead of contributing to global warming as a greenhouse gas, some CO_2 becomes a valuable material to help make better concrete.

Wide-ranging factors are important to developing new green technologies. The most important is the shift in leadership in global innovation dynamics. Although green cement production initially began in developed countries, China is catching up quickly as it is a leader in the global cement market, both as a user and a producer (China's domestic production levels have quadrupled since 2000).

As with many green building materials, regulation is a large factor in adopting new green cement technology. Landlords and contractors in Dubai must now use green cement for new buildings, as a strategy to protect the environment and reduce greenhouse gas emissions through mandating materials that generate less pollution during manufacture. New legislation initiated in Dubai in April 2015 requires all landlords, contractors and consultancy companies to use this environmentally friendly substance before gaining construction permits, and the law provides fines for violations.[22]

Ceramics Used with Cement in Green Buildings

Ceramics will likely become much more widely used as an additive in cement-based composites, incorporating techniques for managing hydration, microstructure, early aging properties, hardening properties, durability, and other physical and chemical phenomena.

Next generation cements will add nano-sized and nano-structured materials, including organic additives and their interactions within cements. This is accompanied by developments in modeling behavioral characteristics such as durability; the interaction of chemical, mechanical and physical factors in cement materials with the environment; hydration kinetics; and microstructural modeling.[23]

Summary

Green design has always been able to get the general public excited. Green design is often glamorous, and it can also be intriguing, compelling, even thrilling to those encountering it for the first time. But its biggest obstacle is a need for greater public awareness and wider understanding of its day-to-day benefits and its role in addressing climate change.

The design community has long recognized the dynamic and accelerating explosion of new building materials, software management tools and technology applications that can enhance our ability to solve sustainability challenges. But building regulations and public acceptance have not always kept pace. Such a broader awareness is essential to effectively support implementing these innovative tools.

By 2020, we expect this transformation to be evident across the entire design and construction stakeholder community, by opening their understanding to the critical need to expand sustainable strategies for bringing advanced green materials into new buildings.

Appendix

2015 LEED Projects Update

To develop the data presented in this book, we used LEED numbers from the public LEED Project Directory. We originally stopped data analysis at the end of 2014, to meet manuscript deadlines, but have been able in many cases to include year-end 2015 information. In this appendix, we provide 2015 full-year results. The overall conclusion: LEED project registrations and certifications have stabilized at 2014 levels, but some project types—schools, universities, retail and healthcare—continue to show sharp declines in new project registrations, indicating a future decline in project certifications.

Nonresidential Projects (US)

1. LEED *Registered* Projects for 2015 are about 10 percent *below* the 2010–2014 annual average but about 7 percent *above* 2014 levels (see Figure 4.1).

2010	2011	2012	2013	2014	Average (2010–2014)	2015
5,653	5,563	4,570	4,100	3,993	4,776	4,257

2. LEED *Certified* Projects for 2015 are 3 percent *below* the 2010–2014 annual average and about 3 percent *below* 2014 levels (see Figure 4.1).

2010	2011	2012	2013	2014	Average (2010–2014)	2015
2,888	3,264	3,501	3,692	3,357	3,340	3,250

3. LEED *Registered* and *Certified* Projects *by area* for 2015 are about 13 percent *above* the 2010–2014 annual average and the 2014 levels (excluding LEED-ND projects and entire college campuses), showing almost no change.

	2010	2011	2012	2013	2014	Average (2010–2014) (million sq. ft.)	2015 (million sq. ft.)
Registered	491	618	610	606	625	590	676
Certified	453	448	387	413	418	424	500

4. LEED-NC/CS *Registered* Projects for 2015 are about 8 percent *below* the 2010–2014 annual average but about 17 percent *above* 2014 levels (see Figure 4.4).

2010	2011	2012	2013	2014	Average (2010–2014)	2015
2,880	3,041	2,739	2,535	2,253	2,690	2,468

5. LEED-EBOM *Registered* Projects for 2015 are about 38 percent *below* the 2010–2014 annual average and about 14 percent *below* the 2014 levels (see Figure 4.4 and Figure 8.8).

2010	2011	2012	2013	2014	Average (2010–2014)	2015
1,750	1,509	787	606	784	1,088	674

6. LEED-CI *Registered* Projects for 2015 are about 12 percent *above* the 2010–2014 annual average and about 17 percent *above* 2014 levels (see Figure 4.4 and Figure 8.9).

2010	2011	2012	2013	2014	Average (2010–2014)	2015
1,023	1,013	1,044	959	956	999	1,115

7. LEED K12 School *Registered* Projects for 2015 are about 28 percent *below* the 2010–2014 annual average and about 16 percent **below** 2014 levels (see Figure 8.3).

2010	2011	2012	2013	2014	Average (2010–2014)	2015
321	246	240	197	207	242	175

8. LEED Higher Education *Registered* Projects for 2015 are about 17 percent *below* the 2010–2014 annual average and about 19 percent *below* 2014 levels (see Figure 8.4).

2010	2011	2012	2013	2014	Average (2010–2014)	2015
540	541	490	464	523	511	422

9. LEED Retail *Registered* Projects for 2015 are about 15 percent *below* the 2010–2014 annual average and about 3 percent *below* 2014 levels (see Figure 8.6).

2010	2011	2012	2013	2014	Average (2010–2014)	2015
1,573	1,480	745	597	929	1,064	904

10. LEED Healthcare *Registered* Projects for 2015 are about 38 percent *below* the 2010–2014 annual average and about 21 percent *below* 2014 levels (see Figure 8.7).

2010	2011	2012	2013	2014	Average (2010–2014)	2015
268	311	190	156	171	219	136

11. LEED-NC/CS *Certified* Projects for 2015 are about 8 percent *below* the 2010–2014 annual average and about 4 percent *below* 2014 levels (see Figure 4.5).

2010	2011	2012	2013	2014	Average (2010–2014)	2015
1,831	2,017	2,214	2,277	1,968	2,061	1,888

12. LEED-EBOM *Certified* Projects for 2015 are about 18 percent *above* the 2010–2014 annual average and about 11 percent *above* 2014 levels (see Figure 4.5 and Figure 8.8).

2010	2011	2012	2013	2014	Average (2010–2014)	2015
439	598	474	505	545	512	604

13. LEED-CI *Certified* Projects for 2015 are about 1 percent *above* the 2010–2014 annual average but about 10 percent *below* 2014 levels (see Figure 4.5 and Figure 8.9).

2010	2011	2012	2013	2014	Average (2010–2014)	2015
618	649	813	910	844	767	758

14. LEED K12 School *Certified* Projects for 2015 are about 7 percent *above* the 2010–2014 annual average and about 6 percent *below* the 2014 levels (see Figure 8.3).

2010	2011	2012	2013	2014	Average (2010–2014)	2015
112	162	206	215	206	180	193 .

15. LEED Higher Education *Certified* Projects for 2015 are about 40 percent *above* the 2010–2014 annual average but about 7 percent *below* 2014 levels (see Figure 8.4).

2010	2011	2012	2013	2014	Average (2010–2014)	2015
161	207	279	342	420	281	393

16. LEED Retail *Certified* Projects for 2015 are about 37 percent *above* the 2010–2014 annual average and about 12 percent *below* 2014 levels (see Figure 8.6).

2010	2011	2012	2013	2014	Average (2010–2014)	2015
188	379	511	611	761	490	671

17. LEED Healthcare *Certified* Projects for 2015 are *about the same* as the 2010–2014 annual average but about 8 percent *below* 2014 levels (see Figure 8.7).

2010	2011	2012	2013	2014	Average (2010–2014)	2015
57	72	110	136	104	96	96

Nonresidential Projects (International)

1. LEED *Registered* Projects for 2015 are about 31 percent *above* the 2010–2014 annual average, about 13 percent *above* 2014 levels and 2 percent *above* 2013 levels, showing slowing growth for LEED in the international sphere since 2013. These 2015 projects represent 35 percent of all newly registered LEED projects globally, about the same percentage as in 2014 (see Figures 4.8 and 4.9).

2010	2011	2012	2013	2014	Average (2010–2014)	2015
1,071	1,684	1,702	2,232	2,018	1,741	2,279

2. LEED *Certified* Projects for 2015 are about 88 percent *above* the 2010–2014 annual average but only about 13 percent *above* 2014 levels, showing slowing growth for LEED certifications in the international sphere during this five-year period. However, since new project registrations *appear to have plateaued* in 2013 to 2015 time frame, one can expect project certifications to plateau also in the 2015 to 2017 period.

These projects represent 27 percent of all newly certified LEED projects globally, compared with 24 percent in 2014 (see Figure 4.9). Certified projects typically show growth from project registrations two to three years in the past, as new construction projects registered in earlier years get completed and then certified. Since most of LEED-registered international projects are for new construction, over time certifications will track at some percentage (approximately 50 to 60 percent) of project registrations from earlier years.

2010	2011	2012	2013	2014	Average (2010–2014)	2015
236	359	583	928	1,049	631	1,183

Bibliography

BRE Global Ltd, 2014, *Digest of BREEAM Assessment Statistics, Volume 01,* available at breeam.com.

Bremmer, Ian, 2015, *Superpower: Three Choices for America's Role in the World,* New York: Portfolio/Penguin.

Bryce, Robert, 2014, *Smaller Faster Lighter Denser Cheaper: How Innovation Keeps Proving the Catastrophists Wrong,* New York: Public Affairs.

Charan, Ram, 2015, *The Attacker's Advantage: Turning Uncertainty into Breakthrough Opportunities,* New York: Public Affairs.

Clifford, Mark L., 2015, *The Greening of Asia: The Business Case for Solving Asia's Environmental Emergency,* New York: Columbia University Press.

Duyshart, Bruce, 2015, *Smarter Buildings. Better Experiences,* Sydney, Australia: bruceduyshart.com/books.

Gottfried, David, 2014, *Explosion Green: One Man's Journey to Green the World's Largest Industry,* New York: Morgan James Publishing.

Hawken, Paul, Amory Lovins and L. Hunter Lovins, 1999, *Natural Capitalism: Creating the Next Industrial Revolution,* Boston: Little, Brown and Company.

Kats, Greg, 2010, *Greening Our Built World: Costs, Benefits and Strategies,* Washington, DC: Island Press.

Kishnani, Nirmal, 2012, *Greening Asia: Emerging Principles for Sustainable Architecture,* Singapore: BCI Asia.

Ma Yansong, 2015, *Shanshui City,* Zürich: Lars Müller Publishers.

Mendler, Sandra F., William Odell and Mary Ann Lazarus, 2010, *The HOK Guidebook to Sustainable Design,* 2nd Ed. New York: Wiley.

McGraw-Hill Construction, 2014a, Smart Market Report: "Green Multifamily and Single Family Homes: Growth in a Recovering Market," construction.com.

McGraw-Hill Construction, 2014b, "Canada Green Building Trends: Benefits Driving the New and Retrofit Market," construction.com.

McGraw-Hill Construction, 2014c, "The Drive Toward Healthier Buildings: The Market Drivers and Impact of Building Design and Construction on Occupant Health, Well-Being and Productivity," construction.com.

7 Group and Bill Reed, 2009, *The Integrative Design Guide to Green Building: Redefining the Practice of Sustainability*, New York: Wiley.

Skopek, Simone and Bob Best, 2015, *Green + Productive Workplace: The Office of the Future...Today*, Chicago: Jones Lang LaSalle.

Yudelson, Jerry, 2013, *The World's Greenest Buildings: Promise vs. Performance in Sustainable Design*, New York and London: Routledge Taylor & Francis.

Yudelson, Jerry, 2011, *Dry Run: Preventing the Next Urban Water Crisis*, Gabriola Island, British Columbia: New Society Publishers.

Yudelson, Jerry, 2009a, *Greening Existing Buildings*, New York: McGraw-Hill.

Yudelson, Jerry, 2009b, *Sustainable Retail Development: New Success Strategies*, Berlin: Springer.

Yudelson, Jerry, 2009c, *Green Building Trends Europe*, Washington, DC: Island Press.

Yudelson, Jerry, 2008a, *Green Building Through Integrated Design*, New York: McGraw-Hill.

Yudelson, Jerry, 2008b, *Choosing Green: The Homebuyer's Guide to Good Green Homes*, Gabriola Island, British Columbia: New Society Publishers.

Yudelson, Jerry, 2007a, *The Green Building Revolution*, Washington, DC: Island Press.

Yudelson, Jerry, 2007b, *Marketing Green Building Services*, New York and London: Architectural Press.

Yudelson, Jerry, 2007c, *Green Building A to Z: Understanding the Language of Green Building*, Gabriola Island, British Columbia: New Society Publishers.

Glossary

ANSI: American National Standards Institute; the organization responsible for accrediting Consensus Bodies that develop national (US) consensus standards.

ASHRAE: American Society of Heating, Refrigerating and Air-Conditioning Engineers; develops ANSI standards that deal with energy efficiency, indoor air quality and thermal comfort. These standards are important references in LEED.

BRE: Building Research Establishment (UK); overseers of the BREEAM system.

BREEAM: BRE Environmental Assessment Method; the green building rating system sponsored by BRE.

EDGE: A global green building certification system for developing economies, promoted by IFC, International Finance Corporation, with certifications provided by both the GBCI and local country organizations.

GBCI: Green Building Certification Institute (relabeled in 2015 as Green Business Certification Inc.); a sister organization to USGBC. GBCI accredits two types of LEED professionals:
LEED AP: LEED Accredited Professional
LEED GA: LEED Green Associate

GBI: Green Building Initiative, Inc.

Green Globes: The building rating system sponsored by the GBI and based on the ANSI/GBI 01-2010 standard.

Green Star: A building rating system sponsored by the Green Building Council of Australia, used in Australia, New Zealand and South Africa.

GRESB: The Global Real Estate Sustainability Benchmark; a rating and reporting system for commercial and corporate real estate, now part of the GBCI.

ILFI: International Living Future Institute; the sponsor of the Living Building Challenge.

LBC: Living Building Challenge, the green building rating system developed by ILFI.

LEED: Leadership in Energy and Environmental Design; the USGBC-created rating system, all rating systems using the LEED label, including:

LEED-NC/BD+C: LEED for New Construction, Building Design and Construction

LEED-CS: LEED for Core and Shell (aimed at new office buildings)

LEED-CI/ID+C: LEED for Commercial Interiors, Interior Design and Construction

LEED-EB/EBOM/O+M: LEED for Existing Buildings, Operations and Maintenance

LEED-ND: LEED for Neighborhood Development.

USGBC: United States Green Building Council.

WELL: A new green building rating system focused on measuring, certifying and monitoring the performance of building features that impact health and well-being, sponsored by the International WELL Building Institute.

Interviews*

Dr. Osman Ahmed, Siemens Buildings Products, Chicago, IL

Bob Best, Executive VP, Jones Lang LaSalle, Chicago, IL

Andrew Burr, Building Technologies Office, US Department of Energy, Washington, DC

Douglas Carney, Senior VP, Children's Hospital of Philadelphia, PA

Larry Clark, Principal, Sustain Florida, Miami, FL

Daniel Davis, Lead Researcher, WeWork, New York, NY

Michael Deane, LEED Fellow, Turner Construction, New York, NY

Gavin Dunn, PhD, Director of BREEAM, Watford, UK

Bruce Duyshart, Principal, Meld Strategies, Sydney, Australia

Charles Eley, Principal, Eley Consulting, San Francisco, CA

Chris Forney, Principal, Brightworks, Portland, OR

Madison Gross, Director, International Council of Shopping Centers, New York, NY

Kimberly Hosken, LEED Fellow, Partner Energy, Los Angeles, CA

Stuart Kaplow, J. D., Stuart D. Kaplow, P. A., Towson, MD

Fulya Kocak, LEED Fellow, Clark Construction, Rockville, MD

Evelyn Lee, Senior Strategist, MKThink, Berkeley, CA

Stefaan Martel, BREEAM Assessor, Bopro, Mechelem, Belgium

Richard Michal, Facilities Director, Butler University, Indianapolis, IN

Rudolph Milian, Senior VP, International Council of Shopping Centers, New York, NY

Brad Miller, Principal, Environmental Concepts Co., Irvine, CA (LEED Project Costs)

* While these people agreed to be interviewed for this book and agreed to the use of any specific quotations attributed to them, they are neither responsible for any part of the contents of this book nor do they specifically endorse any of the analyses or conclusions in the book. Two of the interviewees, as noted, were asked only for information about LEED project costs.

Hernando Miranda, PE, Principal, Soltierra, Inc., Dana Point, CA

Kelsey Mullen, Carlsbad, CA

David Pogue, Global Head of Sustainability, CB Richard Ellis, San Jose, CA

Alan Scott, LEED Fellow, Principal, YRG Sustainability Consultants, Portland, OR

Paul Shahriari, CEO, ecomedes, Atlanta, GA

Kim Shinn, LEED Fellow, Principal, TLC Engineering for Architecture, Nashville, TN

Drew Shula, Principal, Verdical Group, Los Angeles, CA (LEED Project Costs)

Vladi Shunturov, CEO, Lucid Design Group, Oakland, CA

Jiri Skopek, Managing Director Sustainability, Jones Lang LaSalle, Toronto, ON, Canada

Simone Skopek, Jones Lang LaSalle, Toronto, ON, Canada

Curtis Slife, Fellow IFMA, Principal, FM Solutions, Phoenix, AZ

Darrell Smith, Director Facilities and Energy, Microsoft, Redmond, WA

Alexandra Sokol, Principal, EnviroDynamix, San Francisco, CA

Terry Swack, CEO, Sustainable Minds, Boston, MA

Acknowledgments

I want to thank the many green building experts, architects, engineers, contractors, manufacturers, building owners, sustainability directors and other professionals who provided written and oral interviews for this book, supplied project information and as a group lead the way in new thinking about green building. Their perspectives helped to refine the book's arguments and improve its recommendations. Above all, green building is a social and communal activity, with thousands of talented professionals sharing information and perspectives for the past 20 years. Those we interviewed are acknowledged individually, under "Interviews."

I offer a special note of thanks to a longtime green building proponent, Pamela Lippe, LEED Fellow, head of e4, Inc., and founder/president of *Earth Day New York* since 1990, for writing the Foreword, and to Timothy Mack, JD, for writing the Epilogue. I also want to thank the staff at New Society Publishers for supporting this book from its inception and for seeing it through to timely publication.

Thanks as well to the many companies, engineers, building owners, developers, architects, architectural photographers and others who generously contributed project information and photos for the book. Thanks also to Heidi Ziegler-Voll, Heidi 5 Studio, for providing the illustrations we created specifically for this book and to Leslie Evans, Evans Design Group, for providing the "Green Menu" graphic.

For research assistance, especially in analyzing the LEED public project database and providing the graphs and charts of LEED projects in this book, I thank Ayush Nandan Vaidya, M.Arch. Also, Aydin A. Tabrizi, a PhD candidate at the University of Kansas, contributed valuable research and analyses of the 2015 CBECS study and earlier LEED energy performance studies. Thanks to Gavin Dunn and Martin Townsend of BRE for providing updated statistics on BREEAM projects and to Stefaan Martel at Bopro in Belgium for providing the insights of a BREEAM Assessor. Thanks also to Kelsey Mullen for sharing his experience with the workings of the LEED for Homes Multi-family system.

A special note of thanks goes to my longtime editorial associate, Gene Hakanson, for conducting the interviews, sourcing all the photos and permissions (a never-ending task) and adding her experienced touch to the manuscript. This is our 11th book together; she has been an invaluable contributor to each work. I also thank Lindsay Baker, Ann Bartz, Rob Cassidy, Professor Alison Kwok, Kelsey Mullen, Susan Piguet, Paul Shahriari, Steven Straus and Ayush Nandan Vaidya, for reviewing the manuscript and providing many valuable suggestions. Thanks also to Jared Piguet for his help in connecting with new voices in sustainable building.

While I am always appreciative of all the reviewers' work, I accept full responsibility for any errors of omission or commission. Thanks also to my wife, Jessica, for indulging the time spent writing this book and for sharing my enthusiasm for green building.

Endnotes

Reinventing Green Building: A Call to Action

1. 2015 year-end results are based on the author's analysis of data in the LEED Project Directory as of December 31, 2015; details are in the Appendix, 2015 LEED Projects Update.
2. igreenbuild.com/cd_1706.aspx, accessed April 26, 2015.
3. economist.com/blogs/economist-explains/2015/04/economist -explains-17, accessed April 26, 2015.
4. wsj.com/articles/uber-valued-at-more-than-50-billion-1438367457, accessed August 2, 2015.
5. nytimes.com/2015/07/23/nyregion/de-blasio-administration-dropping -plan-for-uber-cap-for-now.html, accessed August 3, 2015.

Chapter 1

1. One of the more astute commentators on society and technology for nearly 50 years, Stewart Brand founded the *Whole Earth Catalog* in the late 1960s. brainyquote.com/quotes/quotes/s/stewartbra172275.html? src=t_new_technology, accessed June 25, 2015.
2. McKinsey Global Institute, "The Internet of Things: Mapping the Value Beyond the Hype," assets.fiercemarkets.com/public/sites/energy /reports/mckinsey-iot-report.pdf, accessed June 30, 2015 and Autodesk, sustainability.autodesk.com/blog/internet-of-things-buildings-aquila/, accessed September 5, 2015.
3. Interview with Dr. Osman Ahmed, Head of Innovation at Building Technologies Division of Siemens Industry, Inc., for Siemens Building Technologies, Inc., June 26, 2015.
4. realcomm.com/advisory/advisory.asp?AdvisoryID=706, accessed August 28, 2015.
5. realcomm.com/advisory/advisory.asp?AdvisoryID=709, accessed September 17, 2015.
6. ecomedes.com, accessed June 22, 2015.
7. gbig.org, accessed June 22, 2015.

8. Charan, 2015, *The Attacker's Advantage*, p. 5.

9. Interview with Paul Shahriari, June 10, 2015.

10. Just as software allows many individuals to file their own income taxes, yet there is still be a need for accountants for more complex returns, in the same way consultants will still be needed to help certify larger and more complex projects.

11. Charan, op. cit., p. 19.

12. forbes.com/sites/genemarcial/2014/04/02/high-frequency-trading -mainly-hurts-the-traders-and-short-term-investors/, accessed September 27, 2015.

13. *Environmental Leader*, a national sustainability newsletter, selected Sustainable Minds Transparency Report™ program as the 2015 "top product of the year," prweb.com/releases/2015/04/prweb12628620.htm, accessed August 19, 2015. See further information on the program at sustainableminds.com.

14. Interview with Terry Swack, June 10, 2015.

15. Bryce, 2014, *Smaller, Faster, Lighter, Denser, Cheaper*, Chapter 23.

16. persistentefficiency.com, accessed June 22, 2015.

17. firstfuel.com/resources/technical-validation/, accessed June 22, 2015.

18. energymanagertoday.com/using-big-data-tackle-big-building-energy -waste-0112413/, accessed June 22, 2015.

19. gocomfy.com, accessed June 22, 2015.

20. fastcoexist.com/3031458/this-smart-new-app-solves-the-number-one -complaint-about-offices, accessed June 22, 2015.

21. Lucid and Urjanet announced in late 2013 a partnership to combine building data with energy data to automate the collection and delivery of utility current billing, history and rate-plan data for Lucid's BuildingOS energy management platform. They teamed up to solve one of the largest barriers to deploying energy efficiency: the cost of acquiring and managing building performance data. luciddesigngroup.com/news/lucid-and -urjanet-team-up-to-enable-universal-energy-data-access-for-building -managers/, accessed September 3, 2015.

22. multihousingnews.com/features/6-game-changers-a-green-building -pioneers-guide-to-disruptive-technology/1004123110.html, accessed July 7, 2015.

Chapter 2

1. Ray Kurzweil is a renowned inventor, recipient of the National Medal of Technology from President Bill Clinton and currently Director of Engineering at Google. Quote is from brainyquote.com/quotes/keywords /trends_2.htmlm accessed June 25, 2015.

2. John Naisbitt, 1982, *Megatrends: Ten Directions Transforming Our Lives*, New York: Warner Books.

3. Patricia Aburdene, 2007, *Megatrends 2010: The Rise of Conscious Capitalism*, Newburyport, MA: Hampton Roads Publishing.

4. imore.com/history-ipad-2010, accessed June 23, 2015.

5. 1.usa.gov/1TLVaCg, accessed June 23, 2015.

6. newbuildings.org/2014-zne-update, accessed June 24, 2015.

7. ecobuildingpulse.com/photos/the-john-w-olver-transit-center and earthtechling.com/2012/10/meet-the-nations-first-net-zero-bus-station/, accessed September 1, 2015.

8. building4change.com/article.jsp?id=2201, accessed June 24, 2015.

9. bomabest.com, accessed April 23, 2015.

10. 2030districts.org/about-2030-districts, accessed April 23, 2015.

11. By mid-2015, a number of large cities such as New York, Philadelphia, Boston and Washington, DC, along with Seattle and San Francisco had such ordinances. blog.wegowise.com/2014-03-27-building-energy -disclosure-laws-the-wegowise-guide, accessed April 23, 2015.

12. A 2013 study by the European Union found positive correlations between a higher rating on the Energy Performance Certificate and higher sales prices and rents in most markets. ec.europa.eu/energy/sites/ener/files /documents/20130619-energy_performance_certificates_in_buildings .pdf, accessed August 12, 2015.

13. Baker & McKenzie, *Global Sustainable Buildings Index 2015*, p. 16, f.datasrvr.com/fr1/015/19281/2015_Global_Sustainable_Building_Index _Full_version.pdf, accessed August 17, 2015.

14. Source: Yudelson, 2009, *Green Building Trends: Europe*, p. 125.

15. epbd-ca.eu/themes/nearly-zero-energy, accessed June 23, 2015.

16. Building Green News, June 22, 2015, www2.buildinggreen.com/dream -materials-webcast.

17. programs.dsireusa.org/system/program, accessed June 23, 2015.

18. cpuc.ca.gov/renewables, accessed June 23, 2015.

19. "The projection of grid parity in the sunniest parts of the United States by 2015 or 2016 appears to have been correct. UBS is informing clients that earlier this year [2015], NextEra, a subsidiary of Xcel energy, submitted bids for new solar projects in New Mexico at a cost of 4.2 cents per kWh. Even after backing out the 30% solar Investment Tax Credit that may soon expire, that would be a cost of 6 cents per kWh, lower than EIA's estimate of 6.6 cents per kWh for new natural gas." rameznaam .com/2011/03/17/expis-moores-law-really-a-fair-comparison-for-solar/, accessed August 17, 2015.

20. google.com/get/sunroof#p=0, accessed August 18, 2015.

21. cleantechnica.com/2015/08/28/mapdwell-moves-san-francisco-finds-3
-gw-potential-rooftop-solar/, accessed August 30, 2015.

22. rameznaam.com/2015/08/10/how-cheap-can-solar-get-very-cheap
-indeed/, accessed August 17, 2015.

23. ww2.kqed.org/science/series/california-drought-watch/, accessed June
23, 2015.

24. Yudelson, 2011, *Dry Run: Preventing the Next Urban Water Crisis*, Chapters 4 and 14.

Chapter 3

1. brainyquote.com/quotes/quotes/m/marktwain122507.html, accessed
June 24, 2015.

2. BRE was privatized in 1997, bre.co.uk/page.jsp?id=1712, accessed April 25,
2015. BRE is now a research charity, and the profits from the BRE Group
businesses are gift-aided to the BRE Trust, who in turn invests in projects
for the public benefit.

3. The author served on the USGBC Board during this period, from 2000
to 2002 and was a LEED National Faculty member from 2001 through
2008.

4. All LEED 2009 projects must register before the October 2016 dead-
line, but will then be given five years (until June 30, 2021) to finish certification.

5. The author was one of the first ten LEED professional trainers, was one
of the first hundred LEED APs certified by USGBC, and personally
trained more than 3,000 building industry professionals through 2008.

6. USGBC 2014 Annual Report, p. 5, readymag.com/usgbc/2014Annual
Report/, accessed December 2, 2015. Many of these accredited profes-
sionals are "legacy" LEED APs who are no longer actively engaged with
LEED projects, but the number is still very impressive, representing
more than twice the current membership of the American Institute of
Architects, for example.

7. gbca.org.au/news/gbca-media-releases/green-star-leaders-recognised-in
-2014/, accessed April 25, 2014.

8. sourceable.net/green-star-statistics-confirm-sustainability-is-here-to
-stay/, accessed August 18, 2015. Since Australia is about 1/15 the size of
the US population, green building certifications there would be equivalent to 3,795 US LEED certifications.

9. canadianconsultingengineer.com/features/history-of-green-building
-rating-systems-in-canada/, accessed April 25, 2015.

10. bomabest.com/about-boma-best/, accessed April 25, 2015.

11. gsa.gov/portal/content/131983, accessed April 25, 2015.

12. The author served as the president of the GBI during 2014 and part of 2015. However, he no longer has any formal or informal affiliation with GBI or Green Globes.

13. usgbc.org/articles/top-10-countries-leed, accessed April 25, 2014.

14. breeam.org/page.jsp?id=347, accessed April 25, 2015. Personal communication, Martin Townsend, BRE, August 20, 2015.

15. Registered project totals include certified projects. Excludes projects outside the United States and also LEED for Homes and LEED for Neighborhood Development projects.

16. worldgbc.org/index.php?cID=220, accessed April 25, 2014.

17. breeam.org/about.jsp?id=66, accessed April 25, 2015.

18. BRE Global Ltd, 2014, pp. 18, 36.

19. Ibid.

20. breeam.org/BREEAMUK2014SchemeDocument/#03_scoringrating _newcon/minimum_standards.htm%3FTocPath%3D3.0%2520Scoring %2520and%2520Rating%2520BREEAM%2520assessed%2520buildings %7C_____2, accessed April 28, 2015.

21. BREEAM, personal communication, Martin Townsend, June 26, 2015.

22. Ibid.

23. usgbc.org/Docs/Archive/General/Docs913.pdf, accessed August 10, 2015.

24. usgbc.org/projects, accessed April 25, 2015. Author's analysis of LEED Project database. Excludes LEED for Homes residential units, all versions. Excludes LEED 1.0 (pilot) projects.

25. Ibid.

26. gbci.org/home.aspx, accessed April 25, 2015. Prior to 2015, it was known as the Green Building Certification Institute, with the same initials.

27. Based on the charts in this book, it appears that one-third of all LEED-registered projects never achieve any level of certification.

28. usgbc.org/articles/introducing-new-leed-online, accessed April 25, 2015.

29. Five new buildings were certified in 2007, including the Arkansas Department of Environmental Quality headquarters in Little Rock. thegbi.org/project-portfolio/certified-building-directory/, accessed August 9, 2015.

30. thegbi.org/green-globes-certification/how-to-certify/, accessed December 5, 2015.

31. Professor Jeffrey L. Beard, "A Study of Comparative Sustainability Certification Costs—Green Rating System Cost Comparison Study: LEED and Green Globes," Department of Construction Management, Drexel

University, March 2014, thegbi.org/training/green-resource-library/30/, accessed June 24, 2015.

32. news.thomasnet.com/imt/2011/06/13/living-buildings-like-leed-on -whole-grain-natural-steroids, accessed June 24, 2015.

33. Twelve of LBC's 20 imperatives are eligible for a conditional assessment after completion of construction, to determine compliance with petal requirements, while eight can be audited only after it is fully completed and operational for at least 12 months: living-future.org/living-building -challenge/certification/certification-details/two-part-certification, accessed April 21, 2015.

34. Yudelson, 2013, *The World's Greenest Buildings*.

35. See for example the seminal 2008 study by New Buildings Institute, newbuildings.org/index.php?q=energy-performance-leed-new -construction-buildings, accessed November 11, 2015.

36. living-future.org/lbc/certification, source: case studies, accessed August 28, 2015.

37. Some will undoubtedly argue that certifications shouldn't be the only gauge of LBC's influence, but if no one actually succeeds in using the full system, how much influence did it really have?

38. phipps.conservatory.org/green-innovation/at-phipps/center-for -sustainable-landscapes, accessed June 24, 2015.

39. Ibid.

40. living-future.org/living-building-challenge/certification/certification -options, accessed April 21, 2015.

41. usglassmag.com/2015/04/living-building-challenge-about-to-get-a -version-older/, accessed April 21, 2015.

42. As noted earlier in the discussion of Green Globes for New Construction as based on an ANSI standard, NGBS is also based on the ICC (Inter-national Code Council) 700 standard. LEED for Homes is a "national" standard, but it is not a consensus standard.

43. homeinnovation.com/services/certification/green_homes/resources /certification_activity_report, accessed January 28, 2016.

44. Personal communication, Kelsey Mullen, former head of LEED for Homes Multifamily program, June 3, 2015.

45. tradingeconomics.com/united-states/housing-starts, accessed June 13, 2015.

46. Yudelson, 2008b, *Choosing Green*.

47. gsa.gov/graphics/ogp/Cert_Sys_Review.pdf, accessed April 21, 2015, at Table ES.5, page xii. The PNNL review studied how the certification systems aligned with the 2011 set of federal high-performance building

requirements using a robustness criterion. There are 27 federal requirements drawn from the Energy Policy Act (1992), Energy Independence
and Security Act, EISA (2007), the High-Performance Sustainable
Building Guiding Principles (2008) and President Obama's Executive
Order 13514 (2009).

Chapter 4

1. The Emperor's New Clothes, a fairy tale by Hans Christian Andersen,
   andersen.sdu.dk/vaerk/hersholt/TheEmperorsNewClothes_e.html,
   accessed Nov. 10, 2015.
2. 2015 year-end numbers are extrapolated from data in the LEED Project
   Directory, as of October 31, 2015.
3. Gottfried, 2014, *Explosion Green*, p. 247.
4. Ibid., p. 250.
5. usgbc.org/projects, accessed August 24, 2015.
6. usgbc.org/articles/leed-certification-update-april-2015, for example,
   accessed May 5, 2015.
7. census.gov/construction/c30, accessed August 4, 2015, using data releases for December 2010 and December 2014. Totals were $687 billion
   in 2014 and $507 billion in 2010.
8. End of 2015 data were projected from numbers in the LEED Project
   Directory as of October 31, 2015.
9. Data are taken directly from the USGBC and GBCI Form 990 (nonprofit) filings with the US Internal Revenue Service. Data for 2014 will be
   available after November 15, 2015. We adjusted 2013 revenues and income
   to account for the one-time sale of a major asset, the *Greenbuild* trade
   show, in 2013, so the chart shows only operating revenues and net operating income. This represents a more accurate indication of the fall-off in
   support for USGBC and LEED.
10. USGBC 2014 Annual Report, pages 5 and 13, readymag.com/usgbc/2014
    AnnualReport/, accessed December 2, 2015.
11. USGBC's annual Form 990 filing to the US Internal Revenue Service
    was not available at the time of writing. The reader can find this on the
    Internet quite easily and can update the numbers. However, USGBC
    reported an operating loss of more than $5 million in its 2014 Annual
    Report, so it's likely that the trends of the previous five years have
    continued.
12. Historically, many of the larger multifamily residential projects have
    used LEED for Core and Shell and LEED for New Construction as the
    certification mechanism, but that changed in recent years as the LEED

for Homes Multifamily certification program gained credibility in the marketplace.

13. census.gov/construction/c30/historical_data.html, accessed May 31, 2015.

14. The calculation is straightforward: LEED-EBOM registered 318 million sq. ft. in 2014, LEED CI 30 million sq. ft., and LEED NC/CS 277 million sq. ft., a total of 625 million sq. ft.; the total built without any project registration was about 1.39 billion sq. ft.

15. usgbc.org/articles/usgbc-announces-international-rankings-top-10 -countries-leed-green-building, accessed August 4, 2015. Author's analysis of data.

16. usgbc.org/articles/leed-homes-international-project-data-update, accessed July 8, 2015.

17. brainyquote.com/quotes/quotes/f/fscottfit100572.html, accessed June 24, 2015.

18. Gottfried, 2014, op. cit., p. 226.

Chapter 5

1. brainyquote.com/quotes/authors/f/francoishenri_pinault.html, accessed June 25, 2015. Pinault is a French business leader.

2. McGraw-Hill, 2014a and b, are recent surveys of the uptake of green building measures by the building industry in the US and Canada.

3. McGraw-Hill, 2014b, p. 46.

4. Edward Mazria, "Turning Down the Global Thermostat," *Metropolis*, October 2003, available at mazria.com/TurningDownTheGlobal Thermos.pdf, accessed June 1, 2015. The figure of 40 percent is often quoted; it might be 38 percent or 39 percent, but the percentage isn't going to change a lot over time.

5. LEEDv4 has a workaround for this prerequisite, but it's unlikely to get much use, according to one experienced consultant I interviewed.

6. Interview with Curtis Slife, May 29, 2015.

7. Interview with Rich Michal, June 29, 2015.

8. See for example, work by Professor Norman Miller, normmiller.net/wp -content/uploads/2012/07/Operations-Management-of-Green-Build ings-Tu-et-al.pdf, accessed September 9, 2015.

9. huffingtonpost.com/roger-platt/the-fourword-business-case_b_792 7616.html, accessed August 5, 2015. A subsequent Platt blog posits that "Our Lenders Value It," huffingtonpost.com/roger-platt/the-business -case-for-gre_b_8185820.html, accessed September 23, 2015, but that

slogan refers only to the same narrow market: large commercial real estate development in big cities.

10. "LEED helps companies show corporate responsibility," by Dawn Killough, greenbuildingelements.com/2015/04/14/leed-helps -companies-show-corporate-responsibility/, accessed April 16, 2015.

11. One question (often) on the survey asks, "Are school buildings that were constructed or underwent major renovations in the past three years LEED certified?" princetonreview.com/college-rankings/green-guide /methodology, accessed June 3, 2015.

12. news.harvard.edu/gazette/story/2015/11/harvard-breaks-leed-record/, accessed December 4, 2015.

13. Author's analysis of LEED public projects database, June 2, 2015. There were 4,706 such institutions, nearly 3,000 of them four-year colleges, in 2011–2012, nces.ed.gov/fastfacts/display.asp?id=84, accessed September 29, 2015.

14. The best source for information on programs and incentives is the Database of State Incentives for Renewables and Efficiency, dsireusa.org, accessed June 3, 2015.

15. President Obama's 2009 Executive Order 13514 required federal agencies to have 15 percent of their buildings over 5,000 sq. ft. meet federal sustainability standards by 2015. In 2015, President Obama issued an even more ambitious executive order to cut greenhouse gas emissions from federal buildings dramatically over the next decade: whitehouse .gov/the-press-office/2015/03/19/executive-order-planning-federal -sustainability-next-decade, accessed June 3, 2015. In 2014, the federal government began a review of appropriate building sustainability standards for the future.

16. earthadvantage.org/assets/documents/AssessingMarketImpactsofThird PartyCertification-090529.pdf, accessed June 3, 2015.

17. earthadvantage.org/assets/documents/EAR%20CommunityDev-Case Study-141110-EOCvs.pdf, accessed June 3, 2015.

18. Information from McGraw-Hill, 2014b, pp. 1, 58.

19. McGraw-Hill, 2014b, p. 1.

20. nreionline.com/office/10-cities-highest-office-rents#slide-10-field _images-1675051 and coydavidson.com/colliers-international/where -are-office-rents-rising-the-fastest/, accessed July 9, 2015. Suburban rents tend to be much lower.

21. One 2008 survey shows this number. usgbc.org/Docs/Archive/General /Docs4111.pdf, accessed June 3, 2015.

22. Assuming $2.50 per sq. ft. for energy, $40 per sq. ft. for rent and $400 per sq. ft. for employees' salaries and benefits (@$80,000 total and 200 sq. ft./employee).

23. Do Green Buildings Make Dollars and Sense? cbre.com/EN/aboutus /MediaCentre/2009/Pages/110209.aspx, accessed June 3, 2015.

24. chgeharvard.org/resource/impact-green-buildings-cognitive-function, accessed November 20, 2015.

Chapter 6

1. Elizabeth Barrett Browning, Sonnets from the Portuguese: 43, poetryfoundation.org/poem/172998, accessed May 16, 2015.

2. For a good, dramatic and very personal accounting of the origins and early successes of USGBC and the LEED system, look at David Gottfried's account in *Explosion Green*, especially Chapter 12, "LEED Rolls Out."

3. Kim Shinn, email survey, June 4, 2015.

4. Interview with Alan Scott, June 5, 2015.

5. aia.org/groups/aia/documents/pdf/aiab089072.pdf, accessed May 18, 2015. By 2008, AIA required four hours of sustainable design education each year in its continuing education requirements, ibid.

6. An eponymous 1990 book about leveraged buyouts popularized the term.

7. Of course, some architects joined this "barbarian horde," including such luminaries as William McDonough (named a "hero of the planet" by *Time* magazine in 1999) at the University of Virginia; Randy Croxton, at the Croxton Collaborative in New York City; and Bob Berkebile at the BNIM firm in Kansas City, Missouri.

8. See for example: Yudelson 2008a; 7 Group and Bill Reed, 2009; and Mendler, Odell and Lazarus, 2010.

9. Renée Cheng, AIA, "Integration at Its Finest: Success in High-Performance Building Design and Project Delivery in the Sector," April 15, 2015, gsa.gov/largedocs/integration_at_its_finest.pdf, accessed July 3, 2015.

10. aia.org/renew/, accessed May 18, 2015.

11. usgbc.org/Docs/Archive/General/Docs3930.pdf, accessed July 3, 2015.

12. *USA Today*'s strong critique of LEED building performance in 2012 especially questioned government tax incentives for LEED-certified buildings, usatoday.com/story/news/nation/2012/10/24/green-building -leed-certification/1650517/, accessed May 18, 2015.

13. E. Mills, et al., "The Cost-Effectiveness of Commissioning New and Existing Commercial Buildings: Lessons from 224 Buildings," evanmills .lbl.gov/pubs/pdf/ncbc_mills_6apr05.pdf, accessed May 18, 2015.

14. Jamie Qualk, "Value of LEED Certification is Third Party Verification," March 2013, facilitiesnet.com/green/article/Value-of-LEED-Certification-is-Third-Party-Verification--13915?source=part, accessed May 24, 2015.

15. For definitions of 108 terms used in green building, see Yudelson, 2007c, *Green Building A to Z: Understanding the Language of Green Building.*

16. Interview with Kimberly Hosken, June 9, 2015.

17. gsa.gov/graphics/pbs/Green_Building_Performance.pdf, accessed May 18, 2015.

18. For a summary of incentives for green building, renewable energy and energy conservation, the best source is still the Database of State Incentives for Renewable Energy and Conservation, dsireusa.org. Also, USGBC's website offers numerous examples: usgbc.org/resources/state-and-local-public-buildings-brief, accessed May 18, 2015.

19. usgbc.org/articles/leading-building-industry-groups-agree-streamline-green-building-tool-coordination-and-deve, accessed May 18, 2015.

20. Writing in *Barron's* financial weekly at the end of 2006, a leading real-estate consultant made this case strongly: "What's going on? A significant real-estate market shift is gathering momentum: Green buildings are going mainstream... To prevent their properties from becoming obsolete, today's real-estate owners should undertake [green] renovations now." barrons.com/articles/SB116683352907658186, accessed August 30, 2015.

21. cbre.com/o/international/AssetLibrary/Green-Building-Adoption-Index.pdf, accessed June 27, 2015.

22. Interview with David Pogue, July 6, 2015.

23. news.investors.com/business-inside-real-estate/062515-759018-small-offices-get-a-black-mark-in-the-area-of-green-buildings.htm, accessed June 27, 2015.

24. David Pogue, op. cit.

25. NGBAI, op. cit., p. 8.

26. NGBAI, op. cit., p. 4.

27. Consider the example of Singapore, where the government's Building and Construction Authority created the Green Mark rating system, requires buildings to use the system, and builds its own demonstration projects, as described in Clifford, 2015, pp. 98–100. Green Mark was updated in 2015, and its report on building energy use shows gains from the requirements: bca.gov.sg/GreenMark/others/BCA_BEBR_Abridged_FA.pdf, accessed September 4, 2015.

28. Academics were able to document this effect as early as 2009. See, for example, Doing Well By Doing Good? An Analysis of the Financial

Performance of Green Office Buildings In the USA, 2009, P. Eichholtz, N. Kok and J. Quigley, srmnetwork.com/pdf/whitepapers/Financial _Performance_of_Green_Office_Buildings_RICS_Mar09.pdf, and The Economics of Green Retrofits, N. Kok, N. Miller and P. Morris, josre.org/wp-content/uploads/2013/01/The_Economics_of-Green _Retrofits-JOSRE_v4-11.pdf, accessed May 18, 2015.

29. This illustration shows how LEED has become such a successful initiative, whereas other systems such as the Green Globes rating system have failed to capture large market share because they lacked the first-mover advantage, the fervent advocates and the comprehensiveness of the LEED ecosystem.

Chapter 7

1. A well-known aphorism, dating to Roman times, in this case attributed to President John F. Kennedy, April 1961, in the wake of the failed Bay of Pigs invasion of Cuba. quora.com/Who-said-success-has-many-fathers -but-failure-is-an-orphan, accessed May 16, 2015.

2. Author's analysis of LEED Project Directory.

3. Interview with Michael Deane, June 8, 2015.

4. Interview with Chris Forney, June 15, 2015.

5. Yudelson 2009a, *Greening Existing Buildings*.

6. usgbc.org/articles/usgbc-announces-extension-leed-2009, accessed September 27, 2015.

7. Interview with Curtis Slife, May 29, 2015.

8. One insider says, "If USGBC improved the certification process and the customer experience was a better one, I believe that LEED would have a much higher adoption rate." LEED has been trying to improve this process for the past six years, without much success, so the adoption rate is unlikely to improve much. In August 2015, I heard from a very experienced LEED consultant based in Virginia that he's been waiting for more than a year for the final project certification, so evidently long waits are not something in the past.

9. See gresb.com, accessed August 25, 2015.

10. Interview with Kimberly Hosken, June 9, 2015.

11. gbic.org/about, accessed May 20, 2015.

12. See also Gottfried, 2014, *Explosion Green*, page 226.

13. eia.gov/consumption/commercial/reports/2012/preliminary/index.cfm, accessed May 20, 2015.

14. Larry Clark, email survey, June 2, 2015.

15. Hernando Miranda, email survey, June 8, 2015.

16. Interview with Douglas Carney, June 5, 2015.
17. Stuart Kaplow, email survey, June 1, 2015.
18. eia.gov/consumption/commercial/reports/2012/buildstock/, accessed January 2016.

Chapter 8

1. Famous phrase first attributed to troubles on the Apollo 13 moon flight, April 14, 1970, brought to life again in the 1995 movie *Apollo 13*, phrases.org .uk/meanings/houston-we-have-a-problem.html, accessed May 24, 2015.
2. The Green Building Initiative's Green Globes rating system had certified nearly 1,000 projects by the end of 2014, LEED nearly 30,000, so the Green Globes share of the total certified market represented about three percent. Counting only projects certified since 2011, when Green Globes incorporated the current ANSI/GBI standard, its current market share is about four percent.
3. Yudelson 2007b, *Marketing Green Building Services*, pp. 165–169.
4. Source: US Energy Information Administration, Office of Consumption and Efficiency Statistics, Form EIA-871A of the 2012 Commercial Buildings Energy Consumption Survey (CBECS), Table 1.
5. eia.gov/consumption/commercial/reports/2012/buildstock/index.cfm, accessed May 24, 2015.
6. 2012 CBECS Preliminary Results, US Energy Information Administration, eia.gov/consumption/commercial/reports/2012/preliminary /index.cfm, accessed May 24, 2015.
7. Ibid.
8. USGBC claims, "88 out of the Fortune 100 companies are already using LEED," leed.usgbc.org/leed.html, accessed May 24, 2015. As with most USGBC claims, this pronouncement must be taken with a grain of salt, since "using LEED" could refer to only one project in a company with hundreds of buildings, and that project might only be registered and not yet certified.
9. edreform.com/2012/04/k-12-facts/#schools, accessed August 15, 2015.
10. LEED for Schools rating system has been around since November 2007, usgbc.org/Docs/Archive/General/Docs2593.pdf, accessed May 24, 2015, and the formal LEED for Schools program launched in 2011.
11. Capital outlay for US schools in 2011 was $56 billion. If each of the new LEED-certified schools had an average cost of $30 million, for example, then 1,044 certified schools costing that $3.1 billion would represent less than six percent of the total amount of new construction and renovation every year.

12. centerforgreenschools.org/newsroom.aspx, accessed May 24, 2015.

13. During a public presentation at a Florida school facilities conference, I heard such a claim in 2014 from the facilities director of a very large school district. The rebuttal is clear: How do you know what was done if you don't document what finally gets built and have a credible third-party review the data? How do you know the schools will save energy unless there's a credible reporting system in place to measure ongoing energy use?

14. census.gov/compendia/statab/2012/tables/12s0278.pdf, accessed June 27, 2015.

15. The author helped create a sustainability master plan for The Ohio State University in 2010. At that time, Ohio State had about 900 buildings and 65,000 students, a ratio of about 70 students per building.

16. Note that LEED for Schools certifies entire schools, whereas LEED certification in Higher Education applies to specific buildings.

17. presidentsclimatecommitment.org, accessed June 27, 2015.

18. usgbc.org/resources/leed-motion-retail, accessed June 27, 2015. Originally released October 23, 2014, usgbc.org/articles/just-released-leed-motion-retail-building-better-global-marketplace, accessed June 27, 2015.

19. Ibid., p. 3.

20. statista.com/statistics/266465/number-of-starbucks-stores-worldwide/, accessed August 9, 2015.

21. en.wikipedia.org/wiki/Verizon_Wireless, accessed August 28, 2015.

22. Half of all certified retail projects used the LEED-CI (commercial interiors) or LEED EBOM (existing buildings) standard, not included in the new construction total.

23. Author's analysis of the LEED Project Directory, usgbc.org/projects, accessed December 2, 2015.

24. statista.com/statistics/266465/number-of-starbucks-stores-worldwide/, accessed June 27, 2015.

25. content.healthaffairs.org/content/34/1/150, accessed August 28, 2015.

26. See Appendix, LEED Project Updates, for the most recent totals.

27. census.gov/construction/nrc/pdf/compann.pdf and usgbc.org/resources/leed-homes-certified-projects (authors analysis) accessed June 24, 2015.

28. usgbc.org/articles/new-usgbc-report-shows-growth-demand-green-homes-150000-now-leed-certified, accessed June 27, 2015.

29. usgbc.org/advocacy/homes-market-brief, accessed August 12, 2015.

30. Information supplied by Kelsey Mullen, former head of LEED multi-family residential program, 2011–2014, August 2015.

31. Ibid.
32. Interview with Kelsey Mullen, June 8, 2015.
33. Individual green raters (certification organizations) tend to do most of the projects in specific states, something one can readily observe in the state-by-state rankings of LEED for Homes projects, indicating they are working closely with multifamily developers in those states. usgbc.org /advocacy/homes-market-brief, accessed August 18, 2015.
34. Interview with Kelsey Mullin, June 8, 2015.
35. Personal communication, Kelsey Mullen, May 2015.

Chapter 9

1. bestofsherlock.com/top-10-sherlock-quotes.htm#incident, accessed June 4, 2015.
2. USGBC and its acolytes will claim, I'm sure, that this shows it has responded to user input by adapting LEED for every market segment. But without changing the overall structure of LEED and the delivery model, our numbers show its use is still shrinking in most segments, as it hasn't convinced any of these user groups (in large numbers) to use the specific adaptations. In this book, I advocate a fundamental rethinking of the entire approach, not just trimming around the edges or boosting requirements as in LEEDv4.
3. Gottfried, 2014, *Explosion Green*, p. 249. (Emphasis added)
4. irs.gov/uac/Newsroom/Filing-Season-Statistics-for-Week-Ending-May -15-2015 and dailyfinance.com/2014/04/14/tax-audit-odds-irs-lowest -years/, accessed June 27, 2015. See also irs.gov/pub/irs-soi/14databk .pdf, accessed December 6, 2015. See also irs.gov/pub/irs-soi/12rswin bulreturnfilings.pdf, accessed December 12, 2015.
5. Interview with Douglas Carney, June 5, 2015.
6. Personal communication, Martin Townsend, head of sustainability at BRE, August 20, 2015 and Stefaan Martel, Bopro, a BREEAM Assessor, August 28, 2015.
7. Personal communication, Kelsey Mullen, July 27, 2015.
8. josre.org/wp-content/uploads/2012/09/Cost_of_LEED_Analysis_of _Construction_Costs-JOSRE_v3-131.pdf, accessed June 5, 2015.
9. Interview with Rich Michal, June 29, 2015.
10. bullittcenter.org/, accessed August 17, 2015.
11. I'm fully aware that many of the "idealists" saw themselves as the true "realists," dealing with the poor energy and environmental performance of buildings. The USGBC itself comprises mostly business members, but they turned over the creation and maintenance of LEED to technical

specialists. Those who were willing to sit "at the table" long enough eventually got their way, as with most systems designed by committees.

12. This can be seen clearly in the area of acceptable forest certification standards: LEED is the only green building rating system in the world that give credits solely to projects that use Forest Stewardship Council (FSC) certified wood.

13. As LEED developed a following among large commercial developers and large corporations, this situation changed quite a bit, e.g., a senior executive of a major developer, Transwestern, served as USGBC board chair in 2013. Several other property developers and managers have served on the board and, with me, on committees such as the one that developed the first LEED for Core and Shell standard. usgbc.org/articles /q-al-skodowski-chair-usgbcs-board-directors, accessed August 30, 2015.

14. Interview with Fulya Kocak, June 30, 2015.

15. huffingtonpost.com/lance-hosey/six-myths-of-sustainable-design_b _6823050.html, accessed June 5, 2015.

16. 100 kWh/m²/year for source energy plus embodied energy is a criterion extensively documented in Yudelson, 2013, *The World's Greenest Buildings*.

17. Interview with Jiri Skopek, June 23, 2015.

18. Interview with Curtis Slife, May 29, 2015.

19. USGBC made it easier in 2015 for projects in California to take this route by aligning LEED requirements with those of CalGreen, usgbc. org/articles/usgbc-announces-new-alignment-calgreen, accessed July 1, 2015.

Chapter 10

1. azlyrics.com/lyrics/kennyrogers/thegambler.html, accessed June 12, 2015.

2. Lisa Matthiessen and Peter Morris, 2007, "Cost of Green Revisited: Re-examining the Feasibility and Cost Impact of Sustainable Design in the Light of Increased Market Adoption," sponsored by Davis Langdon, usgbc.org/resources/cost-green-revisited, accessed August 4, 2015.

3. usgbc.org/Docs/Archive/General/Docs6435.pdf, accessed August 4, 2015; prnewswire.com/news-releases/new-study-finds-green-construc tion-is-major-us-economic-driver-300143364.html, accessed September 16, 2015.

4. Assumes 100,000 sq. ft. NC/CS project, 150,000 sq. ft. EBOM project, and 50,000 sq. ft. CI project.

5. Assumes 550 of projects are EBOM, 850 are CI and 1,950 are NC/CS,

reflecting approximately the number of 2014 US certified projects shown in Table 4.2.

6. Consulting fee of $40,000 represents a consensus of four experienced West Coast consultants in May 2015.

7. Excludes at least 100 hours of staff time; value of at least $10,000.

8. Assumes $0.50 per sq. ft. and 120,000 sq. ft. average size of NC/CS projects.

9. Commissioning is not required in LEED-EBOM but is worth six credit points. Cost of required ASHRAE "Level 1" Energy Audit by a professional engineer would be $10,000.

10. Assumes this is the first LEED-EBOM project for a building owner, so some level of professional services may be required for various analyses.

11. Non-member fees with combined design/construction review, usgbc.org/cert-guide/fees, accessed June 4, 2015.

Chapter 11

1. There are many sources for this quintessentially American saying; here's one: the-working-man.com/alligator-jokes.html, accessed May 25, 2015.

2. builderonline.com/building/usgbc-extends-leed-2009-deadline_0, accessed May 25, 2015.

3. usgbc.org/articles/usgbc-announces-extension-leed-2009, accessed May 25, 2015. For this reason, we can expect a massive influx of new LEED project registrations in the second and third quarters of 2016 and a subsequent very slow start to the LEEDv4 system in the fourth quarter of 2016 and throughout 2017.

4. usgbc.org/LEED/, accessed May 25, 2015.

5. Ibid.

6. Interview with Chris Forney, June 15, 2015.

7. Author's analysis of LEED Public Directory. Current through October 2015.

8. usgbc.org/v4, accessed May 25, 2015.

9. Personal communication, Hernando Miranda, Soltierra, Inc., May 25, 2015.

10. Personal communication, Martin Townsend, BREEAM, op. cit. and Stefaan Martel, Bopro, op. cit.

11. Author's experience with Green Globes as president of the GBI.

12. Schools and Healthcare only.

13. There are no water prerequisites in LEED 2009.

14. Healthcare only.

15. Schools only.

16. Personal communication, Brad Miller, Environmental Concepts Company, May 2015.

17. See 2015 LEED Project Update in the Appendix.

Chapter 12

1. brainyquote.com/quotes/quotes/j/jeffbezos449998.html, accessed June 26, 2015.

2. "2014 Green Building Market Barometer," tinyurl.com/pfvy9fh, accessed August 12, 2015.

3. Interview with Bruce Duyshart, July 8, 2015; also see Duyshart, *Smarter Buildings. Better Experiences*, 2015.

4. A green building certification system is really nothing but a process that, like software, compares a building's design or performance with a set of standards.

5. cbsnews.com/news/chances-of-irs-tax-audit-are-lowest-in-years/, accessed August 31, 2015. See also irs.gov/pub/irs-soi/14databk.pdf, accessed December 6, 2015.

6. nytimes.com/2015/05/12/business/airbnb-grows-to-a-million-rooms-and-hotel-rivals-are-quiet-for-now.html?_r=0, accessed December 6, 2015.

7. energystar.gov/buildings/about-us/facts-and-stats, accessed April 27, 2015.

8. energystar.gov/buildings/about-us/energy-star-certification, accessed June 6, 2015.

9. For the number of GGPs, the source is GBI's *2015 Mid-Year Report*, July 2015, thegbi.org/about-gbi/press-room/article/2015-mid-year-report, accessed September 29, 2015. For LEED accredited professionals, see USGBC's 2014 Annual report, readymag.com/usgbc/2014Annual Report/, accessed December 3, 2015.

10. However, GBI also offers a Guiding Principles Compliance (GPC) assessment protocol, which offers government agencies an opportunity to certify that they are in compliance with sustainability principles developed by a federal Interagency Sustainability Working Group. The GPC compliance assessment tool is easier, faster and cheaper to use.

11. Green Globes still provides credits for doing such things as energy modeling, building commissioning, etc., which are also found in LEED.

12. Updating the 2010 standard, *ANSI/GBI 01-2010, Green Building Assessment Protocol for Commercial Buildings*, began in September 2014 and is expected to conclude by the middle of 2016. Crafting a new Green Globes rating system from the standard, with all of the supporting features, is likely to take well into 2016.

13. brainyquote.com/quotes/quotes/r/robertbrow108884.html, accessed August 4, 2015.

14. Living Building Challenge 3.0, living-future.org/sites/default/files /reports/FINAL%20LBC%203_0_WebOptimized_low.pdf, page 1, accessed April 24, 2015.

15. Interview with Charley Eley, September 8, 2015. The author of the Zero Energy Performance Index (Chapter 15), Eley makes the point that most urban zero net energy buildings cannot have a floor-to-area ratio (FAR) much above 5 to 1 (five stories), because they only have one roof to capture all the solar energy they need.

16. At six (very narrow) stories, Seattle's Bullitt Center is the exception that proves the rule, with a measured annual energy use (before renewables contribution) of about half that of most commercial highly energy-efficient buildings built today.

17. Global average; earthobservatory.nasa.gov/Features/EnergyBalance /page4.php, accessed May 30, 2015.

18. In Los Angeles, for example, from 1999 through 2009, annual rainfall varied from 3.2 inches to 38 inches, a factor of 12 (Yudelson, 2011, page 6).

19. LBC 3.0, op. cit., page 29.

20. See the new study, "Saving Water Increases Health Risks in Green Buildings," rsc.org/chemistryworld/2015/11/water-saving-green-buildings -bacteria-health-risks, accessed December 3, 2015. Abstract can be found at: pubs.rsc.org/en/Content/ArticleLanding/2016/EW/C5EW00221D #!divAbstract, accessed December 3, 2015.

21. ag.arizona.edu/pubs/garden/mg/soils/caliche.html, accessed September 1, 2015.

22. breeaminuse.breeam.org/Home/LogIn?ReturnUrl=/, accessed June 13, 2015.

23. See the BIU Technical Manual for details on what is included in each assessment option, breeam.org/filelibrary/Technical%20Manuals /SD221_BREEAM-In-Use-International-Technical-Manual_V0.pdf, accessed June 13, 2015.

24. Interview with Simone Skopek, July 2, 2015.

25. Quote is from a large corporate client; information taken from a JLL presentation on G+P, courtesy of Bob Best.

26. icsc.org/uploads/about/Property-Efficiency-Scorecard-Brochure.pdf, accessed June 30, 2015.

27. Interview with Madison Gross and Rudolph Milian, July 2015.

28. For example, the Collaborative for High Performance Schools (CHPS) offers an assessment system for K12 schools, which has considerable

practical and policy popularity. CHPS currently offers recognition for school projects in 13 states, including California, Texas and New York. chps.net/dev/Drupal/recognition, accessed June 28, 2015.

Chapter 13

1. Alternative translation of: "One withstands the invasion of armies; one does not withstand the invasion of ideas," en.wikiquote.org/wiki/Victor _Hugo, accessed June 19, 2015.

2. See Yudelson, 2011, *Dry Run: Preventing the Next Urban Water Crisis*, pages 39–56, for an unusually prescient presentation and analysis of multiple and interconnected pathways to permanent water conservation.

3. Source: Yudelson, 2011, op. cit., p. 49.

4. Excluding, of course, nonprofits operating under ANSI consensus standards such as ASHRAE and ICC.

5. aiacc.org/wp-content/uploads/2014/03/AIACC-2013-GBC-Require ments-NonresidentialMandatoryMeasures.pdf, accessed June 14, 2015.

6. pacenow.org/, accessed August 4, 2015.

7. cityofchicago.org/content/dam/city/depts/bldgs/general/Green Permit/110112Green_Permit_Flow_Chart.pdf, accessed June 12, 2015.

8. A good summary for government policy actions can be found in aceee. org/sites/default/files/pdf/fact-sheet/local-govt-ee-policy.pdf, accessed June 14, 2015.

9. 2030districts.org/about-2030-districts, accessed June 14, 2015.

10. Seattle, Los Angeles, Pittsburgh, Cleveland, Denver, San Francisco, Dallas, Toronto, Albuquerque and Stamford, Connecticut.

11. imt.org/resources/detail/map-u.s.-building-benchmarking-policies, accessed June 30, 2015.

12. The American academic Kathryn Janda, currently a senior researcher at Oxford University in the UK, was one of the first to popularize and study this idea, researchgate.net/publication/233501311_Buildings_don%27t _use_energy_people_do, accessed June 14, 2015.

13. Estimates of potential savings range from 2 percent to 10 percent or more. luciddesigngroup.com/blogs/building-occupant-engagement/, accessed June 14, 2015. See also an interesting demonstration project at Fort Carson, Colorado, pnnl.gov/main/publications/external/technical _reports/PNNL-22824.pdf, accessed June 14, 2015.

14. Interview with Andrew Burr, June 16, 2015.

15. See Roger Platt's blog: "Our Lender Values It," huffingtonpost.com /roger-platt/the-business-case-for-gre_b_8185820.html, accessed November 30, 2015.

Chapter 14

1. stephencovey.com/7habits/7habits-habit2.php, accessed June 17, 2015. This is "Habit 2" of the 7 Habits and "is based on imagination—the ability to envision in your mind what you cannot at present see with your eyes." While Covey's book is more than 25 years old (first published in 1989), the principles it contains are timeless.
2. nytimes.com/2015/06/19/world/europe/pope-francis-in-sweeping-encyclical-calls-for-swift-action-on-climate-change.html?_r=0 and nytimes.com/news/un-general-assembly/2014/09/24/ban-ki-moon-says-u-n-must-answer-the-call-to-fight-climate-change-ebola-and-extremism/, accessed June 19, 2015.
3. architecture2030.org/2030_challenges/2030-challenge/, accessed June 17, 2015; 2030 Challenge goals are specific to building type and consider both site and source Energy Use Intensity (EUI); see the following source for details: architecture2030.org/files/2030_Challenge_Targets_National.pdf, accessed June 17, 2015.
4. architecture2030.org/2030_challenges/2030-challenge/, accessed June 1, 2015.
5. nrel.gov/sustainable_nrel/pdfs/44586.pdf, accessed June 14, 2015.
6. energy.gov/eere/buildings/articles/doe-releases-common-definition-zero-energy-buildings-campuses-and, accessed September 17, 2015.
7. energy.gov/sites/prod/files/2015/09/f26/A%20Common%20Definition%20for%20Zero%20Energy%20Buildings.pdf, at page 8, accessed September 17, 2015.
8. Yudelson, 2013, *The World's Greenest Buildings*, Chapter 8.
9. Kishnani, 2012, Greening Asia, pp. 302–319; Yudelson, 2013, op. cit., pp. 191–193. Also, Stephen Wittkopf, "Tropical Net Zero," *High Performing Buildings*, Spring 2015, hpbmagazine.org/Case-Studies/Zero-Energy-Building--BCA-Academy-Singapore/, accessed July 12, 2015.
10. bizjournals.com/sanjose/feature/structures/2014/09/2014/09/best-reuse-rehab-project-winner-435-indio-way.html, accessed December 6, 2015.
11. Solar City had completed more than 200,000 installations as of August 2015. solarcity.com, accessed August 19, 2015.
12. tinyurl.com/hbljpl3, accessed February 5, 2016.
13. Study cited in Yudelson, 2013, op. cit., p. 6.
14. theenergycollective.com/ggkp/2231286/decarbonizing-development-secure-sustainable-future, accessed May 26, 2015.
15. Rob Murchison and Tom Shircliff, "A Beginner's Guide to Intelligent Buildings," RealcommEDGE, May 2015, p. 42, realcomm.com/realcomm-edge/, accessed June 20, 2015.

16. Ibid., p. 43.

17. staples.com/sbd/cre/marketing/easy-on-the-planet/identifying-green
-products.html and officedepot.com/a/guides/buygreen/buygreen/,
accessed June 20, 2015.

18. Carbon emission calculations and reporting for tenant spaces are explicit
in the Green + Productive Workplace rating system described in Chapter
12 and are specifically tied to criteria of the Carbon Disclosure Project,
cdp.net/en-US/Pages/HomePage.aspx, accessed July 2, 2015.

19. Interview with Vladi Shunturov, June 30, 2015.

20. lucidconnects.com/solutions, accessed September 2, 2015.

21. statista.com/statistics/276623/number-of-apps-available-in-leading-app
-stores/, accessed June 17, 2015.

22. leeduser.com/blogs/usgbc-rolls-out-leed-dynamic-plaque-amid-debate,
accessed September 30, 2015.

23. Based on estimates from several insiders with direct knowledge of the
project.

24. greenbiz.com/article/can-leed-go-green-gatekeeper-building-data
-powerhouse, accessed July 11, 2015.

25. From a talk given by Vladi Shunturov at the annual IBCON conference,
San Antonio, TX, June 2015.

26. According to one knowledgeable insider, the price can vary according to
the user, but it is clearly above the cost of much more capable platforms
already in the market.

27. Lucid information comes from Vladi Shunturov, CEO; LDP information
from internal USGBC and industry sources. For the LDP, visit leedon.io
/faq.html.

28. microsoft.com/en-us/stories/88acres/88-acres-how-microsoft-quietly
-built-the-city-of-the-future-chapter-1.aspx, accessed June 19, 2015.

29. Interview with Darrell Smith, Microsoft, July 2015.

30. "Energy-Smart Buildings: Demonstrating how information technology can cut energy use and costs of real estate portfolios," Accenture,
accenture.com/SiteCollectionDocuments/PDF/Accenture-Energy
-Smart-Buildings.pdf, accessed June 17, 2015.

31. Interview with Vladi Shunturov, CEO of Lucid, June 30, 2015.

Chapter 15

1. quotessays.com/reinvent.html, accessed June 17, 2015.

2. Quoted in James Heskett, "Are Technology Companies Ripe for Disruption?" April 8, 2015, Hbswk.hbs.edu/item/7514.html, accessed May 11,
2015.

3. The *Razr* had sold more than 50 million phones by the summer of 2006,

yet was obsolete before the end of 2008. digitaltrends.com/mobile
/ghosts-of-christmas-past-the-original-motorola-razr/, accessed June 19,
2015.

4. Heskett, op. cit.

5. Yudelson, 2013, *The World's Greenest Buildings*.

6. co2now.org/Current-CO2/CO2-Trend/acceleration-of-atmospheric
-co2.html, accessed September 1, 2015.

7. Charles Eley, et al., *Rethinking Percent Savings: The Problem with Percent
Savings and zEPI: The New Scale for a Net Zero Energy Future*, 2011, eley
.com/sites/default/files/pdfs/ASHRAE-D-ML-11-029-20110922.pdf, accessed July 2, 2015 and New Buildings Institute, *Zero Energy Performance
Index (zEPI)*, newbuildings.org//zero-energy-performance-index-zepi,
accessed September 2, 2015.

8. Interview with Charles Eley, September 8, 2015.

9. Eley and others don't want to change the scale; for example, Architecture
2030 plans to continue to state its energy use goals relative to the 2003
CBECS survey, as a result of a 2007 agreement with AIA, ASHRAE
and USGBC to support Architecture 2030. (Personal communication,
Vincent Martinez, Architecture 2030, July 14, 2015.)

10. A term originally coined by the German solar architect Rolf Disch, who
trademarked the term in the 1990s for net-positive homes he designed.
Innovation & Energy magazine (Germany), March 2010, p. 5. See also
plusenergiehaus.de and rolfdisch.de, as well as the article, "Energie
erzeugende Gebäude sollen Standard werden. Plusenergie jetzt!,"
("Energy-producing buildings should become the standard. Plus-energy
now!") in the German magazine *Green Building*, March 2011, accessed
May 28, 2015.

11. Yudelson, 2013, *Dry Run: Preventing the Next Urban Water Crisis*.

12. epa.gov/wastes/nonhaz/municipal/, accessed September 1, 2015.

13. There are many ways to do this. In one shopping mall I visited in Austria
in 2008, bagged solid waste from each store is weighed and the store is
charged directly for all waste they generate. This has the effect of driving
the waste choice upstream to the store's vendors, with an overall beneficial effect on waste reduction practices. See Yudelson, 2009, *Sustainable
Retail Development*.

14. ifc.org/wps/wcm/connect/Topics_Ext_Content/IFC_External
_Corporate_Site/EDGE/Building+typologies/, accessed September 28,
2015. gbci.org/gbci-launches-edge-green-building-certification-system,
accessed December 4, 2015.

15. ifc.org/wps/wcm/connect/Topics_Ext_Content/IFC_External
_Corporate_Site/EDGE/FAQs/, accessed September 28, 2015.

16. LEEDv4 EBOM, EA Prerequisite: Minimum Energy Performance, Case 2, Option 2.

17. whitehouse.gov/the-press-office/2015/03/19/executive-order-planning -federal-sustainability-next-decade, accessed May 28, 2015.

18. Interview with David Pogue, July 6, 2015.

Chapter 16

1. brainyquote.com/quotes/quotes/r/robertfros101324.html, accessed June 17, 2015.

2. brainyquote.com/quotes/quotes/y/yogiberra105761.html, accessed May 20, 2015.

3. See for example, the seminal 1991 book by Peter Schwartz, *The Art of the Long View*, (New York: Doubleday) and the contemporary book by Ian Bremmer, *Superpower: Three Choices for America's Role in the World*, 2015 (New York: Penguin/Portfolio).

4. gbci.org/certification, accessed May 20, 2015.

5. Green Rating for Integrated Habitat Assessment, grihaindia.org, and Indian GBC, economictimes.indiatimes.com/wealth/real-estate/news /india-registers-3-billion-sq-ft-green-building-footprint-igbc/article show/47373537.cms?prtpage=1, accessed May 20, 2015. The Indian GBC reported registering more than 3 billion square feet for green building as of May 2015 and predicted 10 billion square feet by 2022.

6. See Clifford, 2015, pp. 98–100, for more on the story of Green Mark, one of the few green building rating systems for existing buildings developed and supported by a government agency, the Singapore Building and Construction Authority, and one that delivers better buildings in significant numbers. Also, see BCA's 10th anniversary report, bca.gov.sg /GreenMark/others/BCA_Green_Mark_10th_Anniversary_Commem orative_Book.pdf, accessed September 4, 2015.

7. builditgreen.org/greenpoint-rated, accessed June 21, 2015.

8. earthcraft.org/who-is-earthcraft/, accessed June 21, 2015.

9. bit.ly/1frozC3, accessed June 21, 2015.

10. Laura Kusisto, "Luxury Apartment Building Boom Fuels Rent Squeeze," *Wall Street Journal*, May 21, 2015, wsj.com/articles/new-luxury-rental -projects-add-to-rent-squeeze-1432114203, accessed June 17, 2015.

11. Occam's razor is a medieval philosophical construct that yields a simple statement much revered by scientists: *Entities should not be multiplied unnecessarily*, often restated as "the simplest solution is usually the best," math.ucr.edu/home/baez/physics/General/occam.html, accessed September 6, 2015.

12. This is the concept behind the LEED Dynamic Plaque, but it does not

change the LEED EBOM rating system at all, so suffers from all the problems of the LEED system. Even if you "put lipstick on a pig" with some technology, it's still a pig, as the saying goes.

13. "We all need a daily check up from the neck up to avoid 'stinkin thinkin,' which ultimately leads to hardening of the attitudes." —Zig Ziglar, motivatingquotes.com/zig.htm, accessed September 5, 2015.

14. forbes.com/sites/danschawbel/2013/12/17/geoffrey-moore-why-crossing -the-chasm-is-still-relevant/, accessed September 2, 2015.

15. Ibid., italics added.

16. carbonwarroom.com/what-we-do/mission-and-vision, accessed June 21, 2015.

17. presidentsclimatecommitment.org/, accessed June 21, 2015.

18. icsc.org/sustainability#scorecard, accessed June 21, 2015.

19. energy.gov/eere/femp/interagency-sustainability-working-group, accessed June 21, 2015.

20. brainyquote.com/quotes/quotes/w/winstonchu103739.html, accessed June 21, 2015.

21. goodreads.com/quotes/13119-you-never-change-things-by-fighting-the -existing-reality-to, accessed June 30, 2015.

Epilogue

1. Timothy C. Mack is an attorney admitted in New York, Washington and the District of Columbia. He has conducted research with Harvard's Kennedy School, the US National Research Council and the US General Accounting Office, later serving as a vice president for research at WPP Ltd. For 12 years with his own firm, he worked for federal clients, foreign governments, Fortune 50 companies, foundations and NGOs on public policy projects. In 2004, Mr. Mack became President and CEO of the World Future Society, the largest foresight-related NGO, with a global reach. His publications run to more than 40 books and articles. In 2014, Mack established AAI Foresight, a consulting firm and publisher of a white paper series in the strategic foresight field. AAI Foresight resides at aaiforesight.com. Mr. Mack can be reached at tcmack333@gmail .com.

2. quoteinvestigator.com/2013/10/20/no-predict/, accessed June 21, 2015.

3. smitherspira.com/news/2011/march/green-cement-to-take-13-percent -of-market-by-2020, accessed June 15, 2015.

4. materia.nl/material/nabasco/, accessed June 15, 2015.

5. idtechex.com/research/reports/smart-windows-and-smart-glass-2014 -2024-technologies-markets-forecasts-000373.asp, accessed November 28, 2015.

6. idtechex.com/research/reports/smart-windows-and-smart-glass-2014 -2024-technologies-markets-forecasts-000373.asp, accessed June 12, 2015.

7. txactive.us/product.html, accessed June 15, 2015.

8. designboom.com/architecture/achim-menges-developes-hygroskin -and-hygroscope-biomimetic-meteorosensitive-pavilions-4-14-2014/, accessed June 15, 2015.

9. ceres.org/resources/reports/power-forward-2.0-how-american -companies-are-setting-clean-energy-targets-and-capturing-greater -business-value, accessed June 14, 2015.

10. Paul Barter and E. T. Borer, "Implementing Microgrids: Controlling Campus, Community Power Generation," csemag.com/purepower, accessed July 10, 2015.

11. investor.kbhome.com/releasedetail.cfm?releaseid=874346, accessed June 15, 2015.

12. teslamotors.com/powerwall, accessed June 15, 2015.

13. hydrogen.energy.gov/fuel_cells.html, accessed June 15, 2015.

14. A recent study showed that the Tesla system has market potential in terms of cost, smartgridnews.com/story/mit-looks-reality-tesla-battery -costs/2015-06-16, accessed June 17, 2015.

15. navigantresearch.com/newsroom/green-building-materials-will-reach -254-billion-in-annual-market-value-by-2020, accessed November 28, 2015.

16. blogs.ei.columbia.edu/2012/05/09/emissions-from-the-cement -industry/, accessed June 14, 2015. epa.gov/ttnchie1/conference/ei13 /ghg/hanle.pdf, accessed June 14, 2015.

17. epa.gov/osw/nonhaz/industrial/special/ckd/rtc/chap-2.pdf, accessed June 14, 2015.

18. forbes.com/sites/jenniferhicks/2014/06/23/green-cement-to-help -reduce-carbon-emissions/, accessed June 14, 2015.

19. statista.com/statistics/248667/size-of-the-global-cement-market/, accessed June 14, 2015.

20. economist.com/news/business/21579844-worlds-cement-giants-look -set-recoverybut-will-it-be-durable-ready-mixed-fortunes, accessed June 14, 2015.

21. carboncure.com, accessed September 30, 2015.

22. citiscope.org/citisignals/2015/new-buildings-dubai-must-use-green -cement, accessed September 30, 2015.

23. ceramics.org/meetings/3rd-advances-in-cement-based-materials-charact erization-processing-modeling-and-sensing, accessed November 28, 2015

Index

About the Author

Dubbed the "Godfather of Green" in 2011 by *Wired* magazine, Jerry Yudelson is the leading published authority on green building and sustainable design. A noted national and international keynote speaker, Jerry has spoken on green building, green homes, water conservation and sustainable urbanism in 17 countries since 2006. In 2011, Jerry became one of the original 34 LEED Fellows named by the USGBC and GBCI.

Jerry's professional career has addressed water, energy and environmental issues, along with the design, construction and operation of green homes and commercial green buildings. He works for architects, developers, builders and manufacturers to facilitate sustainable design solutions.

Besides his extensive business and professional background, Jerry served eight years as an original LEED national faculty member for the US Green Building Council (USGBC). Beginning in 2001, he trained more than 3,500 building industry professionals in the LEED green building rating system. He served on the USGBC's national board of directors and, from 2004 through 2009, chaired the steering committee for USGBC's annual conference, Greenbuild—the largest green building conference in the country.

In 2014 and part of 2015, Jerry served as president of the Green Building Initiative, a national nonprofit organization with a mission to accelerate the adoption of green building.

Jerry Yudelson is the author of 13 prior green building books, including *The World's Greenest Buildings: Promise vs. Performance in Sustainable Design*; *Dry Run: Preventing the Next Urban Water Crisis*; *Greening Existing Buildings*; *Green Building Through Integrated Design*; *Green Building Trends: Europe*; *Sustainable Retail Development: New Success Strategies*; *Green Building A to Z: Understanding the Language of Green Building*; *The Green Building Revolution*; *Choosing Green: The Homebuyer's Guide to Good Green Homes*; and *Marketing Green Building Services: Strategies for Success*.

Jerry Yudelson founded Yudelson Associates in 2006, an international green building and sustainability consultancy. He holds undergraduate and graduate degrees in civil and water resources engineering from the California Institute of Technology (Caltech) and Harvard University, as well as an MBA (with highest honors) from the University of Oregon. He is a professional engineer and has been named as a National Peer Professional by the US General Services Administration (GSA).

Jerry and his family live near the Pacific Ocean in Oceanside, California.

If you have enjoyed *Reinventing Green Building*, you might also enjoy other

Books to Build a New Society

Our books provide positive solutions for people who
want to make a difference. We specialize in:

**Climate Change ◆ Conscious Community
Conservation & Ecology ◆ Cultural Critique
Education & Parenting ◆ Energy ◆ Food & Gardening
Health & Wellness ◆ Modern Homesteading & Farming
New Economies ◆ Progressive Leadership ◆ Resilience
Social Responsibility ◆ Sustainable Building & Design**

For a full list of NSP's titles, please call 1-800-567-6772 or check out our web site at:

www.newsociety.com